NAVAL WAR GAMES

NAVAL WAR GAMES

World War I and World War II

BARRY J. CARTER

ARCO PUBLISHING COMPANY, INC.
NEW YORK

For Sylvia

Published by Arco Publishing Company, Inc.
219 Park Avenue South, New York, NY 10003

Library of Congress Catalog Card Number 74-24664
ISBN 0-668-03748-2

Printed in Great Britain

CONTENTS

LIST OF ILLUSTRATIONS

PLATES

IN TEXT

The photographs reproduced in this volume are by Iradj Parvaneh. With the exception of 3, 4 and 12, the figures are by Michael F. Childs.

INTRODUCTION

They are only models and the sea may be any surface which is conveniently to hand, but when the grey warships slip out of port at the start of a war game they become something different. They are German cruisers heading for the Atlantic to prey upon enemy shipping; they are the battleships, carriers and escorts of Force H leaving Gibraltar for another desperate voyage escorting a convoy to Malta in World War II; or they are an American carrier task-force off on a long probe towards the Japanese-held islands of the Pacific. It does not matter that rulers are being used for measuring ranges or that dice will be rattled for measuring the accuracy of the gunfire; for the player concerned his ships are at war.

From now on he nurses his forces, keeping a watchful eye on the expenditure of fuel and ammunition and biding his time before throwing in his ships in order to wrest the initiative from his opponent. At another time he will give ground in order to gather his strength for a decisive blow elsewhere. He watches with satisfaction as his opponent's ships reel out of line damaged or sinking, or with scarcely concealed impatience as an enemy ship moves unsuspectingly into range of his lurking submarine. Or he launches his aircraft only to see their attack repulsed and his carrier put in jeopardy from enemy attack as it waits for the returning aircraft. If this battle is only part of a longer campaign the player will weigh up the probable effect on the other theatres of war if he should fight aggressively and take chances on the one hand, or should avoid contact with the enemy for as long as possible on the other. Decisions such as these will keep him alert and involved for the duration of the battle and go a long way towards explaining the fascination of naval war gaming. But this is only part of the appeal of the hobby.

Before the stage has been reached of pitting ships against each other, the players will spend many hours in preparation. First comes the selection of a suitable venue for the battle, with an appropriate set of circumstances to explain why the two countries are at war. Then the ships to be involved are purchased or built from scratch. If commercial models are chosen, then the players make their selection from a bewildering choice of scales, manufacturers and materials; if they build some for themselves, this involves a great deal of pleasurable study of reference books to establish the models' precise measurements, disposition of superstructure and colour scheme. Rules are then framed to take account of ships' speeds, armour and armaments in order to achieve reasonable realism in the coming battle. Thus by the time the war game is due to begin, the players concerned are steeped in the subject and watch over their model fleets with very real concern. And when the fighting is over and the ships are packed away again, the players mull over the outcome and speculate how the rules may be improved for the next encounter or a new situation be contrived for another battle.

Between times the players are almost certainly adding to their knowledge of modern naval warfare and of war gaming in a variety of ways. They may belong to a war-game club which meets regularly at a player's home or in a hired hall in order to fight more battles or to hear guest speakers. Sometimes the club will hire a feature film or entertain the members from another club. Ships and books will be sold or exchanged, and above all ideas and experiences are interchanged. Occasionally, too, the naval war gamer will have the opportunity to visit a national convention where battles are fought for prizes, where manufacturers and publishers show their latest wares and where like-minded people once again pass on tips to each other.

Even the lone war gamer can keep in touch with developments in the hobby by taking one or more of the increasing number of journals catering for his interest in naval battles in miniature. Eventually he is almost certain to discover an opponent against whom to pit his wits—probably by seeing a personal advertisement in one such journal.

Should the war gamer find he has exhausted the possibilities for one particular historical period, he can turn to another and begin once again reading, researching and building up new fleets. Or he may experiment with different scales finding that really large models give a personalised form of warfare whilst by contrast microships allow whole campaigns to be waged in a confined space. For a complete change he can, in any case, turn to board games which allow for all the excitement of battles with model ships without the corresponding expenditure of time and money in preparation. War gamers with models and those with board-game equipment tend to choose one but not the other, although there is no reason why this should necessarily be so since both approaches allow one to simulate naval warfare.

Another dimension is given to the hobby by the kind of activities it engenders. The enthusiast finds himself visiting maritime museums and specialist libraries, the former for helpful models, dioramas and general inspiration, the latter for a key reference book or a rare book. Preserved warships provide inspiration, so too does the sight of a ship with which one has actually modelled or battled. The war gamer seeks out feature films and documentaries relating to his period of interest, exchanges letters with other war gamers, and above all makes many lasting friendships.

In the chapters which follow I try to help the war gamer in the modern period by describing all those facets of the hobby which present difficulties to the newcomer and even the experienced player. Commercial models are described and their principal suppliers named. Some advice for building ships from scratch is given. Popular and readily available sets of rules are assessed, and then the author's own set of rules are proffered for use, adaptation or even rejection! Some actions using these rules are then described, and these may suggest other scenarios for readers to set up. Chapter 6 points the way to finding out more about the hobby, and the final chapter describes the best-known makes of naval board games. Finally the notes to these chapters and the appendices offer the reader guidance in the enjoyable process of becoming more experienced in his hobby.

I am sometimes asked if I do not find the whole subject of war gaming distasteful. Certainly one cannot read a book about naval warfare without being reminded of the suffering, waste and boredom associated with the occasional drama and heroism that appeal to the gamer. But I regard any form of history as a legitimate form of study, and it would be surprising in this century if the two world wars had not made a deep impression on many of us. Moreover, it has not been my experience that those people who recreate warfare in miniature are any less constructive members of society than those who do not. It is more likely that war gamers are people who already enjoy competitive games such as chess, draughts, bridge or even Monopoly. A more serious charge is that war gaming strengthens those tendencies in most societies to resolve disputes by force. This indeed may be the case, although it is equally possible that playing at war provides an outlet for some of our aggressive tendencies. In the last resort our naval battles are games and no one has ever left one in which I have been involved with worse than a headache or injured pride!

B. J. C.

Chapter 1

CHOOSING MODEL SHIPS

The most popular scale for models for warship battles on the table top or on the floor is 1/1,200 and its continental counterpart, 1/1,250. The former is the International Ship Recognition Scale and indicates that 1 inch on the model represents 100ft on the real ship. A County Class cruiser thus becomes 6·3in on the war game ocean, and the battleship HMS *Duke of York* becomes 7·45in. Continental models are slightly shorter, but when models of the two scales are mixed the result is certainly no worse than the mixing of 1/76 and 1/87 armoured fighting vehicles. Some reference books used by ship modellers, for example Blundell's *Royal Navy Warships* and the German Weyer's *Taschenbuch der Kriegs Flotten*, draw all their plans to 1/1,200 scale.

It would be pleasant to go on and say that the reader need only go into any reputable model shop and start buying his ship models in either scale. But neither in Britain nor the United States has this been the case for some years, and generally the newcomer to the hobby has to order by post using advertisements in the modelling and war-gaming journals. However, the increase in the numbers of firms producing war-game ships recently suggests that naval war gaming may once again enjoy the popularity it did in the 1950s when both Eagle and Minic ships were widely available in British shops.

To help the newcomer, here is a list of all the firms who make or who have made ships in the recent past and which the war gamer might expect to come across.

MODELS IN 1/1,200 SCALE

AIRFIX The announcement in 1973 that this firm was going to start producing a range of war-game ships was welcome news indeed. Its other war-game and modelling products enjoy a wide sale in Britain and abroad, and can be obtained from a great many shops. The firm intends to produce a series of British and German warships of the World War II era with the more famous ones given preference—the first to appear were *Bismarck* and *Hood*. Airfix kits are plastic, well produced and very reasonably priced.

EAGLE Eaglewall produced their series of detailed plastic kits in the late 1950s and early 1960s, all neatly boxed with an assembly sheet carrying a lengthy description of the actions in which the ship of the kit had taken part. Most helpful of all was the fact that the ships were issued in Battle Series so that one could always expect to have opponents for each model. Although the eighth and last series was designated 'The Battle of Cape Matapan', the scheduled Italian vessels never appeared; thus Eagle ships are confined to British and German vessels of World War II.

As Eaglewall ships started so many of us off in naval war gaming and so many enthusiasts are still seeking them out, it is perhaps worth giving more precise details of the kits issued. But it should be borne in mind that the names on the boxes are a little misleading as they suggest that the models are of particular warships, whereas one kit was serving for all the ships of a class. Nevertheless the range of ships was very extensive and enabled one to build up interesting fleets—not omitting less glamorous types of vessel such as depot ships and trawlers. Certainly one could refight the actions of the Battle of the River Plate, the capture of the *Altmark*, the Battle of Narvik, the Battle of the Atlantic, the sinking of the *Bismarck*, and the sinking of the *Scharnhorst* (using the *Gneisenau* kit).

The ships produced were as follows:

British

Queen Elizabeth class battleship
King George V class battleship
County class cruiser
Leander class cruiser
HMS *Exeter* cruiser
Illustrious class aircraft carrier
H class destroyer
K class destroyer
Tribal class destroyer
Black Swan class frigate
Captain class destroyer escort
Castle class corvette
Tree class trawler

A tanker appeared in various boxes as the German tankers *Altmark* or *Nordmark* or as a British oil tanker.

German

Bismarck battleship
Gneisenau battlecruiser
Graf Spee, *Admiral Scheer*, *Lutzow* } pocket battleships—each kit different
Hipper type cruiser
Von Roeder class destroyer
Maass class destroyer
Saar submarine depot ship
Acheron U-boat tender
U-boats of types IX, VII, II, IXc and XXI

Gneisenau was produced in very small quantities.

Although Eaglewall ceased production in 1962, their ships are still to be found occasionally in model shops and sometimes appear advertised second-hand in war-game magazines. Astonishingly, Egon Wielding of Munich still advertises consignments of the unmade kits from time to time. I understand that Lone Star models of the United States have held the ship moulds since Eaglewall's demise.

ENSIGN These were welcome newcomers in 1973. The Ulster manufacturer has achieved the ideal balance between the detail possible in a metal model and a price the average war gamer can afford. Smaller vessels only have been produced so far, and these are for World War II. British and German submarines predominate, but the range also includes a Narvik class destroyer, British Hunt class destroyer escort, and the rarely modelled German F class escort vessel. The makers also produce accessories to the same scale, some of which enable one to provide land-based features for the naval war gamer to attack, for example, lorries, railway engines and wagons.

FLEETLINE Highly detailed and accurate metal models in kit form or painted. Models on sale so far include the World War II British escort carrier *Attacker*, destroyers *Armada* and *Jutland* and the submarine *Tally-Ho!* Individual parts such as hulls, torpedo tubes and guns can be supplied separately. The firm also buys and sells other makes of 1/1,200 scale ships and harbour accessories.

GILLOW SHIPS It must be stressed that Mr Gillow is a one-man producer of metal warships, and therefore his range of models cannot be compared in scope or quality with those of established firms. But he has usefully concentrated on the smaller vessels like destroyers, corvettes, gunboats and motor torpedo boats, so often neglected by the big firms or priced too expensively to enable one to build up flotillas. His models are only loosely based upon World War II prototypes; indeed this approximation is intentional so that the war gamer can make conversions or designate a model as any vessel he wishes. The ships are quite nicely cast and come either painted, or unpainted for half the price. Turrets do not move and detail is rather lacking, but at the very low price one can scarcely complain.

LEICESTER MICROMODELS Another recent and welcome newcomer to the war-gaming field, Leicester Micromodels, are going to concentrate on the smaller vessels of the World War II period—probably not going above the cruiser in size. The ships are of metal and come unpainted. The detail is good especially on the motor torpedo boats. The first models to appear have been the British monitor *Roberts*, a Town class destroyer, a Tribal class destroyer, an A class submarine, a Flower class corvette and a D class Fairmile motor torpedo boat. The German ships are a Narvik class destroyer, a torpedo boat and a 740 ton U-boat.

LINDBERG LINE Correspondence with the makers confirms that there are only four models and that they are not to a constant scale. Indeed it is the standard size of the packaging which appears to have decided the size of each ship. Each kit includes a full hull, a display stand and decals, and each is described on the packet as being 'more than six inches long'. Happily, however, the USN cruiser *Biloxi* is just about correct to the 1/1,200 scale and with the hull cut down to the waterline is a pleasant addition to any Allied fleet.

MERCURY A small range of ships, distributed on the continent and mostly mercantile. But they include the Italian cruiser *Montecuccoli*, which was completed in 1934. It has a metal hull with plastic turrets.

MISCELLANEOUS In the context of war games, this almost invariably means 'made in Hong Kong' because a number of plastic waterline or full-hull models emanate from the Crown Colony. It takes a keen eye for bargains to spot such ships, but a clue I would offer is that one should look in fancy-goods shops rather than in model shops—they frequently provide a source for merchant vessels which one needs for convoy battles. It is a curious fact that a Minotaur class cruiser is always turning up from Hong Kong, and it is so like the Triang Minic model that I can only conclude a straight copy has been made. Modern aircraft carriers also turn up occasionally as do Battle class destroyers.

MINISHIPS These plastic kits from Italy were first sold in Britain in 1972 and there are now a number of distributors for them in several countries. The 1,200 Model Ship Society sells them to members at a discount.

Each model arrives assembled and finished in dark grey. The various parts can be unplugged for detailed painting before permanent assembly using polystyrene cement. Transfers for the ship's pendant numbers are included as are adequate assembly instructions and some hints on achieving a camouflage effect when painting. A commendable feature is that the three aircraft carriers of the same class—*Enterprise*, *Yorktown* and *Hornet*—incorporate differences which were present in the actual ships. The *Enterprise* for example carries more anti-aircraft guns than *Yorktown* which was sunk early in the Pacific War. Then again the elevators giving access for aircraft are placed in different positions of openness in all three. A final feature which is not found in many firms' aircraft-carrier models, is the large number of aircraft supplied for each ship. *Yorktown* has twenty-three, of which fifteen are on the flight deck, and as all are removable one can transfer some to other firms' models. *Enterprise* has fewer aircraft, whilst *Hornet* comes with the B25 bombers which enabled the carrier to strike at Tokyo in the 'Doolittle Raid' in 1942.

In addition to these three models, the following World War II vessels have so far appeared: the battleships *Iowa*, *Wisconsin*, *New Jersey*, *Missouri*, *Yamato*, *Musashi*, *Roma*, *Littorio*, *Vittorio Veneto* and *Impero*, and the Japanese carrier *Shinano*. There are apparently plans to produce the World War II vessels *Bismarck*, *Tirpitz*, *Jean Bart*, *Richelieu*, *King George V*, *Prince of Wales*, *Duke of York*, *Ark Royal* and a number of postwar ships. These kits are good value, my only criticism being that they all seem to show too much of the hull for ships presumably fully laden.

PYRO These plastic kits were made by Pyro Plastics Corporation of New Jersey, USA. No additions have been made to the small range for some years, and it is believed that the firm has recently discontinued their production altogether. Some Pyro products now appear as Life-Like Hobby Kits, but not this range of ships. Early sales literature listed some major ships, including cruisers, of Britain, the United States, Germany and Japan and these were obtainable in the United States. In Britain, however, only eight differently named kits appeared, and sadly one has to report that these eight kits only represent four warship types. These are battleships of the South Dakota and Yamato classes, and aircraft carriers of the Essex and Zuikaku classes.

RECOGNITION MODELS (British Admiralty) Since these elusive wooden waterline models are sometimes advertised in British hobby journals, it is worthwhile explaining something about them. They are produced by top-class model makers for use as recognition models in the Royal Navy and they come on to the market whenever their real ship equivalents have been scrapped. Shops specialising in ex-Admiralty clothing and equipment tend to acquire such rejected models. The majority of the ships I have come across are post-1945, with a great many of the Soviet navy and other rarities such as ships of the Swedish and Netherlands navies. But vessels like *Jean Bart*, *North Carolina*, *Atlanta*, *Gambia*, and Fletcher and Battle class destroyers can all be pressed into World War II service.

SUPERIOR This New Jersey company's large range of World War I and World War II vessels are readily obtainable in the United States and becoming increasingly available in Britain.

For World War II there are more than 200 different ships

of the 1939–45 fleets of the USA, Britain, France, Germany, Japan and Italy. There are some notable omissions—particularly of British carriers—but compensations such as the inclusion of auxiliary vessels and the choice, where a ship was substantially reconstructed at any time, of a model of the original ship or the reconstructed vessel. The World War I range is smaller but continues to grow.

The models are metal castings and are spray-painted grey with details picked out in appropriate colours. All gun turrets of 6in and above revolve as do cranes. Recently a number of models have been reissued with improvements such as the separate casting of secondary guns.

For war gamers who, like myself, connive at a *Vanguard* appearing in World War II battles, the range includes the United States battleship *Montana* (which was, in fact, never built) with its twelve 16in guns.

The main source of Superior Ships is by buying direct from the makers, Alnavco of New Jersey, USA. They offer a choice between the finished model and a kit which costs up to 30 per cent less; the completion of the kit involves some work with pliers, a file and paints.

Alnavco also make aircraft in 1/1,200 scale—generally five to a packet—featuring the Corsair, Avenger, Hellcat, Helldiver, Zero, Val and Kate types.

Superior's products compare favourably in price with German models of metal construction.

TRIANG MINIC Long before the Triang company's bankruptcy and dispersal in 1971–2, it had been decided to discontinue the Minic range of waterline warships and merchant vessels, all of metal construction. Happily one still comes across both new and second-hand items in the range of ships and harbour accessories, an Ulster firm being particularly successful at securing warships in mint condition.

In all some twenty liners were issued including the *Queen Mary* and *Queen Elizabeth* and these can be legitimately used in war games of the World War II period as fast troop ships which was their wartime role. The eleven miscellaneous vessels included a whale-factory ship, which can be pressed into service as a depot ship or part of a convoy, whale chasers (submarine chasers), a paddle-steamer (estuary anti-aircraft vessel) and a tug. Then there were twenty-eight pieces for making up a harbour: breakwaters, quays, storage tanks, piers, dock gates and so on. All could be placed on green plastic sea which came in two sizes, the larger being 52in square. There was even a whale and the Statue of Liberty!

The eleven warships were all of the Royal Navy and mostly postwar, but the cruisers *Swiftsure* and *Superb* and the unmodified A class submarine can legitimately be used in late-war actions. I also press into service the modified Battle class destroyers and the last British battleship, *Vanguard*. Of 42,500 tons displacement, the *Vanguard* carried eight 15in, sixteen 5·25in and seventy-one 40mm guns, and incorporated many of the lessons learned during the war, especially the need for numerous anti-aircraft guns. But although laid down and launched during the war, her completion was not hurried once the Allied victory was seen to be inevitable. Instead she acted in peacetime as a training ship and was used extensively to show the flag, much as the *Hood* had done between the wars. We were never to know how effective she would have been in battle conditions, but it is fun to try and see in a table-top game.

MODELS IN 1/1,250 SCALE

If comprehensive fleets are to be built up, there is little doubt that one will sometimes mix models of different scales and in the case of the 1/1,250 range, the difference from the 1/1,200 models is comparatively slight.

It is on the continent that the 1/1,250 scale holds sway, and most of the ships described below can be obtained from any of three German retailers although there are a growing number of outlets for them in Britain and the United States. The models are generally made of metal and depending on the accuracy and detail reproduced, are rather expensive and will deter many war gamers unless they are backed by a deep war chest!

ANKER A small range of expensive models and primarily of the World War II period. Turkey and Yugoslavia are represented, whilst the uncompleted Italian aircraft carrier *Sparviero* is an unexpected item. So is the rarely modelled British depot ship *Maidstone* which has recently achieved new prominence as a prison ship for political detainees in Northern Ireland.

ATLANTIC Again, a small range and consisting mainly of recent passenger ships, tankers and freighters. Two, *Jan Molsen* and *Vaterland*, could legitimately be used in World War II games.

COPY Twenty or so British, French and Italian warships, including some like *Enterprise* and *Arethusa* which are not often modelled. Reasonable prices.

DELPHIN This firm produces more than sixty ships which could be used in war games of World War II, and of most combatant navies, including the Netherlands. The models are completely made-up, in metal and are painted. Well made but expensive.

HANSA Makers of warships, merchant ships, aircraft and harbour accessories, all to the same scale. This firm is particularly strong on American cruisers and German destroyers and torpedo boats, whilst there is a model of the German aircraft carrier *Graf Zeppelin* as she would have looked had she been commissioned. The quality of these metal models is high but so is the price. A substantial saving is possible if they are bought in white metal kit form. There are more than 200 ships available mostly of the World War II period.

MERCATOR Again, a very large choice of ships in metal and painted is available. As the majority are ironclads and merchant vessels of the period between 1880 and World War I, this is a good source for those wishing to build up a fleet for the early period. There are some unusual models for the later period such as the Japanese Kaiten-type midget submarines and Japanese aircraft-carrying submarines.

NAVIS This firm offers a magnificent selection of ships, also primarily of the World War I period. German and British ships are particularly well represented but sufficient are available for Japan, the United States, France, Italy and Austro-Hungary to make many scenarios possible. The models are in metal with plastic masts, derricks and turrets.

NEPTUN Virtually all the Japanese navy's major warships which took part in World War II are produced by this firm, together with a few destroyers and a tanker. The rest of the range is entirely of American warships, mostly destroyers. Beautifully detailed models.

OCEANIC The British producer of these highly detailed models has concentrated on providing types which other manufacturers have tended to neglect. Thus motor torpedo boats, trawlers, tenders, tugs and numerous auxiliaries predominate in a range that now numbers about 150 types. The majority of these are of World War II British warships. The models are in metal and can be purchased fully assembled and painted, or in unpainted kit form.

SCHLINGELHOF Primarily makers of plastic aircraft in the same scale and therefore very useful for adding to models, they also produce a handful of warship kits in plastic, including the American escort carrier *Casablanca*, a tank-landing craft and and American Elco PT-boat.

STAR This Austrian firm concentrates on the smaller vessels of both World Wars but has a selection of ironclads from the American Civil War. In kit form the models cost about 75 per cent less than in the finished form. Many countries are represented and there are some curiosities such as the Chinese cruiser *Ping Hai*, the German armed merchant cruiser *Coronel* and the British monitor *Roberts*.

TRIDENT A very large range from the 1920s to the 1960s is offered with Japanese ships of World War II particularly well represented. A number of merchant vessels come in kit form.

WIKING This German firm's products are more readily available outside Germany than many others mentioned above. But the range is confined to the World War II and the postwar period and is not very extensive. Plastic is used for the more fragile parts of superstructures, but the capital ships come mostly in metal. Seven countries are represented. The capital ships include some interesting types—*Renown, Nelson, Marat* (of the Soviet Union), *Italia* and *New Mexico*. The smaller ships include the Norwegian minelayer *Tryggvason* and destroyer *Sleipner*, and the British fast minelayer *Manxman*.

SMALLER SCALES

Although most war gamers enjoy battling with ships in the scales I have dealt with above, it must be admitted that the playing surface needs to be quite large to give the movements and firing any degree of realism. And a large amount of space is not often convenient to retain for more than a few hours in the average household. Thus models in the scale of 1/3,000 make it possible to fight actions on smaller surfaces or to have larger engagements realistically on table top or floor space.

The following British firms are all introducing models in the smaller scales.

SKYTREX (1/3,000) Made of plastic resin, each ship comes individually packed with specification and undercoated ready for painting. All those issued so far are of World War I vintage and feature the principal British and German Dreadnoughts, battlecruisers and some representative cruisers and destroyers.

NAVWAR (1/3,000) A recent series, again in plastic resin. The first few for 1939–45 provide the starting point for refighting the *Bismarck* action in 1941 being the *Bismarck* herself, *Prinz Eugen*, *Hood*, *Prince of Wales*, *Rodney* and a County class cruiser. Although beyond the scope of this book, it is perhaps worth mentioning that Navwar also have an extensive range of pre-Dreadnought vessels for Britain, Japan, Russia and Germany.

MICROSHIPS (1/4,800) Smaller again—now we really can refight Jutland on the tabletop! So far the World War I range has been confined to British and German vessels and mostly of the size of cruiser and above. But for World War II there are more than fifty vessels for Britain, Germany, Japan, the United States, Italy and France. The models are somewhat crudely finished and lack all but the basic elements of hull, superstructure and main turrets, but provide a more attractive 'counter' than any cardboard symbol used in table-top gaming.

MINIFIGS (about 1/2,000) This enterprising firm (Miniature Figurines) of soldier makers has a small range of British World War I ships which are modestly priced but lacking in accuracy.

AND GOING LARGER ...

The next popular scale above 1/1,200 is 1/600 and the British firm of Airfix make several British and German warships, primarily for World War II, whilst Revell in the United States and a number of Japanese firms make models of the same size or larger. Most war gamers will feel they cannot use such large models, however, except for occasional ship-versus-ship encounters or at a war-game gathering where space is not restricted. Also ruled out for the same reason are the excellent kits in 1/700 scale put out by the Japanese firms of Aoshima, Fujimi, Hasegawa and Tamiya. In time they expect between them to produce all of Japan's 1941–5 navy. Several of the destroyers are easily converted to cruisers in 1/1,200 scale.

Yet another scale became available in Britain with the arrival of Saito waterline warships from Japan at the end of 1973. These are plastic kits in 1/1,000 scale and the first ones are for the Japanese warships *Shokaku*, *Zuikaku*, *Yugiri*, *Fubiki*, *Akizuki* and *Haruzuki*. The carriers have two submarines with each kit. The last-named vessels look quite like light cruisers when used with models in 1/1,200 or 1/1,250 scale.

Chapter 2

BUILDING FROM SCRATCH

Those readers who quail at the thought of trying to produce their own war-game ships can take heart from the advice that follows since I must disclaim any special expertise in the matter. I began to make my own or convert existing models when ready-made types were either unobtainable or too expensive. I have no particular facility for making or repairing things. That said it must be admitted that making one's own ships is a very satisfying task and the saving in cost is so great as to open up all sorts of possibilities for building up fleets in different periods and for different countries.

MAKING A SHIP

HULL TEMPLATE When the ship to be modelled has been chosen, it is necessary to find a plan giving adequate detail for copying purposes, preferably with photographs which convey the essential style of the vessel. A reference book like *Jane's Fighting Ships* offers both, of course, but to model in the 1/1,200 scale it is wise to have access to a drawing in that scale and Jane's plans are not constant in this respect. Weyer's *Taschenbuch der Kriegs Flotten* does have ship plans to this scale, so do the inexpensive and growing range of books by the British publishers Almark. A template should now be drawn for the hull by tracing the plan on to tissue paper and thence on to a piece of stiff card which can be cut out. If the chosen plan is not of the required scale, then a free-hand sketch will often do just as well.

THE HULL I use balsa wood for most of my modelling and this can be bought in various thicknesses and lengths or in bundles of assorted sizes. A sheet of about $\frac{1}{4}$in thickness is adequate for the hull of most battleships, but thinner sheets are necessary for smaller vessels. An aircraft carrier, with its great height, is best cut out of a block of balsa or built up with several thin layers. In each case the template is placed on the balsa sheet and fastened with a pin at bow and stern for extra security. A razor blade or sharp-bladed modelling knife is then used to cut out the wooden hull. A certain amount of smoothing with a very fine glass-paper may be necessary at this stage. If you are making more than one ship of a class it is sensible to make all the hulls at the same time.

THE SUPERSTRUCTURE For the larger items of the superstructure, pieces of balsa wood may again be used and reference to the plan will indicate what sizes and shapes are necessary. But already the pieces of wood being handled are quite small and balsa splinters or squashes very easily. So for a finer finish it is best to use pieces of plastic sheet which, like balsa wood, can easily be obtained from model shops. These sheets come in various thicknesses—the thinner kinds being flexible enough to allow curved shapes to be made for, say, a funnel cowling. Layers built up like a sandwich may provide a more realistic-looking bridge than strips of balsa wood sliced with difficulty and perhaps inaccurately. A word of warning about adhesives. Balsa cement provides excellent adhesion for two surfaces of balsa as does polystyrene cement for plastic surfaces. But if balsa is joined to plastic, or indeed if any other unlike materials need to be joined, it is wise to use a multi-purpose clear adhesive. Turrets are best left untouched until their fixing agent is thoroughly set, since the placing of gun barrels can sometimes unseal a turret which is not firmly stuck down.

Funnels can look difficult to model, but any good model shop catering for the aircraft enthusiast will have lengths of round dowelling for sale. Several lengths of differing diameters can be chosen and a section cut from the most suitable. If an oval-shaped funnel is required, then light sandpapering will give the required shape.

GUNS AND MASTS Fuse wire is as satisfactory for these as any other material. Wire that will be thick enough to look like large calibre guns, such as 15in or 16in will probably have to be cut with a special wire-cutting tool. For secondary guns, a razor blade or knife should be adequate for cutting the necessary lengths. Philatelists' tweezers are useful to manoeuvre these tiny pieces of wire into position. If your turrets are made from balsa wood, it is not too difficult to push the wire guns into place although a spot of all-purpose glue put on both turret and gun makes for a permanent join. Masts can be shaped and glued in the same fashion, the mast becoming very secure if its base is forced into the wood a little way.

THE TRIMMINGS One's model should now have the profile of the real thing but will look very bare and decidedly wooden. Here is the moment to dip into the spares box every modeller should have by him in which to pop unwanted items from other models, and there are several suppliers of such accessories in 1/1,200 scale. The ship will really start to look like a warship if torpedo tubes are added—more pieces of wire will do for these, together with lifeboats, life-rafts, searchlights, anti-aircraft guns and range finders. If the vessel carries a plane and catapult, then the former can be bought ready-made whilst the latter need only be a sliver of balsa wood. Finally the modeller should tidy up the model by carefully removing stray strands of polystyrene cement using a razor blade and tweezers.

PAINTING Balsa wood needs to be coated with a transparent primer before it will take paint, and since this dries quickly it is possible to give the vessel two or three coats in an evening. It does not matter if the non-balsa fittings get coated at the same time. The overall colour for most vessels in wartime is dark grey but there are some variations from this and the modeller should consult his reference books for the correct shade to use. To relieve the greyness one can paint funnel tops, the mouths of gun barrels and torpedoes matt black, and the tops of lifeboats and life-rafts matt white. If the vessel is one of a flotilla a coloured band may be painted round a funnel, and if it is the flotilla leader two bands will distinguish it from the others. A dot on the funnel side will serve the same purpose. I am assuming the reader will use paint-brushes although spray paints are available and are a quick if somewhat expensive means of applying the overall colour. Experience will decide the modeller's choice of paints. Very extensive ranges in gloss and matt are sold and come in tins or bottles.

It is fun to camouflage your vessel, but you should not do this if you are already happy with a model's finished appearance. After all the camouflage will only serve as decoration in the war game. There are numerous photos of ships in camouflage rig in books such as the Purnell Histories of the World Wars, and a work has recently been published solely on this subject and dealing with British warships (see notes to this chapter). The essential feature of any camouflage scheme is that it should break up the lines of the vessel so that an enemy observer has difficulty in determining the ship's speed, course and precise profile. In World War I a 'dazzle' effect was produced by painting lines over the hull and superstructure. But more subtle colour and line schemes were produced in World War II, a favourite ploy being to paint a bow wave some way aft with the hope of confusing the gun and torpedo aimers.

Finally you can add any suitably small flags which come your way. Very often these will be the smallest ones out of larger-scale kits as for many years no manufacturers gave any attention to providing such accessories for small ships. Since 1972, however, it has been possible to obtain in German model shops sheets of flags 6in by 4in, each sheet having more than one hundred items. The various sheets give flags and admirals' ensigns for all the combatant nations in the two wars, international signal flags, funnel stripes and 2mm numbers, and funnel markings for some of the world's shipping lines.

NAMING THE MODEL If you have been building a model for which there was only one original, perhaps HMS *Hood*, it is a simple matter to finish the model by painting the name on the bottom of the hull. Even so it is useful to put a number in the same place, too, giving a different one to every ship in your fleet. Then if you have a battle in which fictitious names are given to the vessels the number will serve to identify the ship and the new name can be written on the Damage Sheet.

You may have built several ships of the same class and you now have to assign names to them. This is a pleasant job and one for which you may adopt several criteria. In the first place you may want to have in your fleet ships whose real counterparts had distinguished war records. For example, *Emden* is a must for the name of a World War I cruiser. Or you may wish to have ships all of which were built at your favourite dockyard. In my own case I tend to choose Devonport-built ships. Or again, and this will be the case with German destroyers particularly, you may use those ships with names and omit those that only have numbers.

CONVERTING EXISTING KITS AND MODELS

With the steady increase in supplies of metal and plastic ships of all the various scales suitable for naval war gaming, the conversion of existing models to fill a gap in a fleet is no longer so necessary. But if I give some examples of conversions that I have made for World War II battles, the reader can judge for himself when it becomes necessary or worthwhile to embark on the occasionally hazardous job of altering a model.

REDUCING FULL-HULL MODELS Packets of plastic full-hull merchant ships are always turning up in the shops and provide a ready source for vessels for convoy actions. The Lindberg line of ships also comes in full-hull form whilst Casadio Miniships, although waterline models, are too high out of the water for realism. In each case surplus hull can be removed by first scoring a neat line with the modelling knife to show where the division is to be made. The knife blade is then held in a flame for a few minutes and applied firmly and quickly along the scored line. The unwanted plastic comes away easily enough. An unheated knife can be used, but brittle plastic will sometimes snap and spoil the hull entirely. Sandpaper is used to smooth the roughened edges of the hull.

USING UNWANTED HULLS The novice modeller will sometimes find hulls difficult to fashion for his scratch-built ships, especially aircraft carriers. It seems to me to be perfectly legitimate to utilise the otherwise useless full hulls that come with such kits as Eaglewall and Pyro for the basis of another ship. And since the Italians never completed their projected carriers *Sparviero* and *Aquila*, I have had no compunction in building them up from the full hulls of Pyro battleship kits. Of course, on close examination, they look rather odd but the racy line of their profile does not seem out of character with other Italian warships. Even more successful in its finished form was the escort carrier USS *Sangamon* whose hull was the unwanted one of a plastic freighter from a ubiquitous Hong Kong packet. One of my earliest conversions of this type was to build up the 1925 cruiser *Emden* on a light cruiser full hull from an Eaglewall kit. It is unsatisfactory when viewed at eye-level, but thanks to the correct layout of the superstructure is a passable substitute for a commercial model.

CONVERTING ONE SHIP TO ANOTHER My fleets include some successful examples of these, but some failures too. The Eaglewall *Exeter* kits proved suitable for making the German light cruisers *Konigsberg*, *Karlsruhe* and *Koln* of the inter-war years, the hull being stuck together unmodified and then cut to reduce its length by half an inch or so. Barrels were added to the turrets to make them triple mountings and they were placed as for the German design. But I was also tempted by having a surplus of Eaglewall County class cruiser kits to try making other cruisers of foreign navies, and these have not worked out well. There is something unique about those great sides of the British 'Washington Conference' cruisers which defy even drastic surgery. So my Japanese cruiser *Mikuma* sports the fifteen 6·1in guns which I was so anxious to get on to a model, but the hull is still unmistakably British. Similarly the Italian cruiser *Bolzano* was easily copied for its superstructure details, but even when severely cut down the County class hull is recognisable. I have been apprehensive about tinkering with metal models—I have heard of too many accidents when modellers have ended up with blobs of melted metal instead of ships—but did try to give the British Royal Yacht *Britannia* a new career as the fast minelayer *Ariadne*. Again, this was not a successful transformation. The metal of Triang Minic ships is resistant to all but savage treatment, and so I ended up with two new and accurate funnels alongside the shapely mercantile funnel of the original. Moreover, the plastic guns and lifeboats with which I festooned her did little to disguise her original lines. She is now relegated to hospital-ship duties.

INCORPORATING DIFFERENCES IN VESSELS OF THE SAME TYPE This is a type of conversion which never fails to interest me and which can provide some fascinating hours of consulting reference books in order to achieve the most realistic results. A large class of vessel which lends itself very well to such attention is the British County class cruiser. The Eaglewall models of two of these, *Dorsetshire* and *Norfolk*, seem to have been manufactured in great quantities and still appear on the market occasionally as unmade kits. As named here, they need not be altered at all but all the rest of the class have significant differences. The most markedly different is the *London* which was taken in hand between 1938 and 1941 and restyled in the manner of the more recent Fiji class of ships. Thus it had two funnels (one higher than the other) instead of three, a more massive bridge, and a revised layout of the superstructure including increased anti-aircraft armament. *Cumberland* and *Suffolk* had increased armour protection and gained aircraft hangars, but to compensate for the increase in displacement the hull was cut down one deck abaft the rear 8in gun turret and the torpedo tubes were removed. *Cornwall* and *Berwick* gained hangars, lost tubes, but retained the deck aft.

In making *Cumberland* and *Suffolk*, I removed the piece of hull in the manner described earlier and cut the same length of decking off the main deck piece so it could be reseated in the lower position. The hangars were made of cardboard, although plastic card would do equally well. Tripod masts, where necessary, were made up of conventional masts with pieces of wire added to give the other two legs. Otherwise the conversions merely involved moving pieces such as cranes, catapults and secondary weapons to different parts of the superstructure.

All navies made alterations and improvements to their vessels so that ships within a class could often differ as markedly as did the County class cruisers. To take British examples again, the five Queen Elizabeth class battleships were taken in hand with various degrees of severity between the wars and whilst all ended up with one funnel instead of two, they differed considerably in appearance by the nature and arrangement of their secondary armament. Dido class anti-aircraft cruisers should have all carried ten 5·25in dual-purpose guns, but wartime shortages led to *Charybdis* and *Scylla* being armed with eight 4·5in guns instead. Such differences are generally not difficult for the war gamer to incorporate in his models, and even if there are no known differences between two vessels there is no harm in altering the position of lifeboats and rafts between one and the other.

Chapter 3

RULES FOR WAR GAMES

The reader has now acquired some warship models and will wish to begin staging battles. If he is already quite knowledgeable about twentieth-century naval warfare, he will be aware of some of the problems which occur when trying to turn real warfare into a simulation. Simultaneous movement by all the ships concerned can scarcely occur unless the war gamer is prepared to devise an electronic war game. The firing of guns and the recording of hits are simple enough to arrange if chance plays a large part in the calculations, much less easy to do if skill is to be exercised. What about limits on fuel and on ammunition? Can weather be built into the game so that heavy seas or fog will upset an admiral's plans? Can credit be given to one nation's possession of the better equipment or another's rigorous training of its sailors? Will a crew's performance depend to some extent on a nation's political and economic health or on the strength of its naval tradition? These are the sorts of questions the war gamer finds himself asking, and for answers he may well turn to one of the commercial sources for a set of rules. This chapter therefore introduces some of the best known or most readily available rules. Such is the healthy state of the hobby that we can expect many more to appear, some linked with a particular firm as a means of promoting their model ships, others as yet another attempt to produce a definitive set of rules.

One whole area of dispute in rule making has been the absence of unchallengeable data on the performance of modern naval guns in terms of range, rapidity of fire and effect of shells. And if some of this data has existed it has been difficult to cull from the various sources. But an increase in the release of such data by maritime governments as the necessary security prohibitions have been relaxed, and the greater availability in published form, means that we can expect to see new attention given to this aspect of rule making. Two war gamers in the forefront in this respect are A. D. Zimm and Andrew J. F. Smith. The former is a member of the Spartan group of clubs in California, the latter of the 1,200 Model Ship Society in Britain. Mr Smith's carefully collated data on the naval guns of many nations has been appearing in the war-game journal *Broadside* (see Chapter 6) whilst Mr Zimm's new rules 'Battle Stations', based on detailed gunnery research, were published in *The Spartan* in 1973 (see Appendix 3, and for an address see notes for Chapter 1).

Of the rules already available, I comment first on those intended solely for actions in the period 1914–18, secondly on those which will suffice for any action in the period 1914–45 and, finally, on those intended for warfare in the period 1939–45. In the case of two well-known naval war gamers of other years, the British Fred Jane and the American Fletcher Pratt, I comment here only on aspects of their rules. A fuller treatment of their contribution to the hobby is given in Chapter 6.

Fred Jane's last set of rules appeared in 1912 and are thus suitable for virtually all actions of the type which occurred in World War I. They are not easily consulted, but it seems likely that they will eventually be reprinted. As this enthusiastic war gamer invited serving naval officers to participate in his battles, he endeavoured to achieve maximum realism even so far as taking account of the working of electric hoists in gun turrets in formulating his rules. Whilst such attention to detail will not appeal to every war gamer, Mr Jane's rules on movement, firing and damage assessment are sensible and rapid in use. Movement is across a board ruled in equal-sized squares, speed being expressed as so many squares per move and in proportion to a vessel's known speed in knots. Ranges of guns are measured in the same fashion. Like all war gamers, Jane had to struggle to produce a satisfactory method of aiming and firing. His solution may appeal to the reader. A pin was placed off centre and projecting downwards from a large piece of circular card. The card in turn was fixed at right angles to a stick. When 'firing' the player lifted the stick its own length above the target and then plunged it down hoping to hit something vital with the pin which was, of course, not visible to him. Damage charts enabled the umpire to assess the effect of the hit. Jane comments on this method:

> It is not possible to aim with the new pattern striker if the firer is much excited; thus, in a roundabout way, some equivalent for 'moral effect' is produced. I have never yet seen a player able to deliver a well-aimed fire after his ship has been badly knocked about. The reason is that he is doubly anxious to hit—just the thing that makes hitting difficult.

Depending on what part of the ship has been hit, turrets were deemed out of action or the ship's speed and manoeuvrability impaired, whilst points were recorded for each hit until sufficient had been amassed to sink the ship.

Again, Jane has some workable suggestions for giving merchant vessels a vital part in the game. In a campaign-type game one could well try his rules for the commerce raider. He regarded ships at sea as passing constantly at *x* hour intervals. The names of these ships, their speed and points value, were placed on cards in a box and one card was drawn out every *x* hours, provided the commerce raider was deemed to be astride their route. The speed of the merchant vessel was then read off. If its speed was greater than that of the commerce raider then it was deemed to have escaped; if slower, time was given for a chase to take place and the destruction or capture of the vessel. The commerce raider was regarded as having a particularly high fuel consumption and its fire-power effectiveness was reduced as prize crews depleted its manpower. If the country promoting the commerce-carrying countered the effect of the raider by employing the convoy system, then it lost game points because of the delays involved in organising the convoy.

Enough has been said to indicate that Fred Jane's rules have a relevance for any naval war gamer and it is a pity that they have been so difficult to consult over the years.

In contrast to Fred Jane's rules, which appeared more than sixty years ago, another set for World War I actions was published in 1973 by John Hammond for Skytrex Limited of Leicestershire, England. The twenty pages of his rules form an attractive booklet, clearly printed with good explanatory diagrams and tables. They are suitable for the period 1914 to 1928 and therefore give full rein to those ships beloved of war gamers—battleships and battlecruisers. Because Skytrex ships are in the 1/3,000 scale so are the measurements of range and speed that Mr Hammond introduces, but they can be converted for use with 1/1,200 scale ships by a simple calculation. A certain amount of equipment is necessary—a tape measure, percentage dice (which generally consist of a pair of multi-sided dice which can produce all the scores between 1 and 100), turning circles made out of card and the stationery usual for any game. No specially marked playing area is necessary, however, since moves and firing are measured with the tape. Sixty per cent of the rule book is devoted to explanation, the remainder to tables showing the ranges of guns and the damage attributable to them expressed in points. The major guns of the warships of Argentina, Austria, Brazil, Britain, Chile, France, Germany, Holland, Italy, Japan, Russia, Spain, Turkey and the United States are tabulated in the latter section and thus save the war gamer much time and effort in research.

In common with most war gamers, Mr Hammond has a formula for assessing a ship's resistance to damage, this being expressed in points which then become the total required before the ship sinks. Damage affects the ship's performance and a simple table lists the results for any vessel. To allow for the magazine explosions which were fairly frequent amongst warships before the experience of the Battle of Jutland led to improvements, extra dice are thrown when a ship has experienced severe damage to ascertain whether the dreaded cordite flash has occurred. Both in terms of this hazard and of armour protection in general, Mr Hammond gives extra credit to German warships whose virtues in this respect are well known.

Most other features of naval warfare are introduced in these rules—torpedoes whether fired from surface vessels or submarines, mines and their clearance, the consumption of fuel and its replenishment, ramming, the weather and the hazards of range-finding in the days when ships produced dense clouds of smoke. Altogether a well-formulated set of rules which provide both realism and a fast-moving game.

Another set of rules designed specifically for World War I actions and again linked with a particular model scale are Keith Robinson's set *The Last Sea Battles* for Leicester Micromodels Limited. This firm makes ships to 1/4,800 scale, and these were described in Chapter 1, but the rules can be adapted for any scale up to 1/1,200. As with Mr Robinson's set for World War II actions, these rules put particular emphasis on the effects of the weather on warship activities. I describe these more fully later in this chapter. Also, there are similarities to the other set in respect of movement, firing and the effect of successful hits on a target; but special attention is given to the organisation of groups of ships and their rigid system of manoeuvring when operating as a fleet, and interruptions of signalling between ships produces exciting results.

For campaign warfare it is suggested that movement be transferred to a map marked in squares or hexagons and the search procedures which then become necessary can be played by two players alone or, better still, with an umpire. These rules come in an attractively produced, nineteen-page booklet.

Victory at Sea is the title of the rules first published in 1971 under the editorship of A. J. Morale with the co-operation of fellow-members of the Nassau Naval and Military Historical Society. This rule manual is well established in the United States but only recently has been used by war gamers in other countries. Twelve pages of the manual are devoted to rules, with examples of their use whilst there are seven appendices as separate charts, mostly giving tables for gunnery ranges, penetration and damage. Some of these are printed on both sides of a sheet, and for quicker consultation in a battle one would need to photocopy these tables so that all seven appendices could be displayed together. It is claimed for the rules that they can be applied from the pre-Dreadnought era to the 1950s, and that an action like that between *Bismarck* and *Hood* and *Prince of Wales* in 1941 can be played in as little as forty-five minutes. Time is an important factor since there is nothing more frustrating than being unable to resolve a battle because each game move has taken so long—and the time taken over a move can be the result of over-complex rules or tardy players. *Victory at Sea* does not seem to have the first fault whilst tardy players can be hurried by regulating moves with a clock as is the case in chess.

Movement is measured with a tape, so is gunfiring, but a protractor is used to calculate a ship's turning circle and to estimate the angle for torpedo firing. Dice and a deck of playing cards complete the equipment. The major section of the manual deals with movement, sighting and smokescreens, firing of guns and torpedoes, submarine and anti-submarine rules, aircraft and anti-aircraft rules and the optional device of salvo firing. Some suggestions for setting up strategic actions combining map moves with battles with models are briefly considered in the second section, and the third section gives three examples of the rules in action.

These examples could, perhaps, have been usefully expanded to supplement the otherwise excellent explanation of each process in the game. The authors have collected a great deal of information on warship performance and have formulated workable rules. They have included one novel technique, that of salvo firing. This is a device already familiar to land war gamers who often make allowance for the spread of shot. It is an optional addition to the rules and its use introduces an element of skill since the player chooses between normal, tight or elongated spreads and then places the measuring device at the spot where he estimates the target to be. The spread of shells is indicated by markings on the plastic spread indicator and produces a more realistic pattern of fire than that used in the conventional methods where shots all fall neatly in the same plane.

Fletcher Pratt, like Fred Jane before him, brought to naval war gaming a deep knowledge of maritime affairs and it is therefore no surprise that the rules he formulated with his New York friends in 1940 have stood the test of time. Many of today's war gamers will have been influenced by his rules and his devices for simulating movement, firing and assessing damage have not, to my mind, been greatly improved upon.

For example, Pratt's Ship Value Card fulfils many functions once the original research and lengthy but simple arithmetic are done. Thus once a ship has been given a points value derived from that ship's speed, tonnage, armour and armaments it is a simple task to assemble opposing fleets of equal strength whilst the number also serves as the ship's sinking value, it being regarded as sunk when more than that number of damage points have been scored against it. For the sake of convenience, Pratt also divided the ship's points value by its speed and the figure derived served as the amount needed for the attacker to reduce the ship's speed by 1 knot. Guns and torpedoes could be knocked out of action at certain set intervals on this points value scale too. Thus a ship having a points value of 30,000 and a speed of 30 knots lost 1 knot in speed for every 1,000 damage points incurred. Guns were knocked out at regular intervals like 27,500 points, 25,000 points, 22,500 points and so on.

To calculate a composite figure for a ship a value for guns is assigned first, the same arithmetic being applied to main, secondary and tertiary guns. Then follow allowances for the number of torpedoes carried, the thickness of armour, the number of mines and aircraft carried and then the whole sum is multiplied by the ship's speed. The global points total is then added to the ship's tonnage to complete the value score.

When ships fire guns and torpedoes under the Pratt rules, players are required to estimate the distance to the target and to allow for the movement of the target. Precise measurements are then made to see if the shells or torpedoes have in fact hit the target. This system replaced a variety of different methods all of which had been tried and found wanting. They included the firing of air pistols at silhouettes of the ships, using pencil-beam torches, toy spring cannons firing wooden plugs and every variation of dice throwing. Altogether one can recommend Fletcher Pratt's rules as being a successful blend of the realistic and the workable.

Of the rules featured in this chapter, Keith Robinson's set *Naval Warfare 1939–45* for Leicester Micromodels Limited are the most helpful for war gamers who want their battles to occur as part of a larger campaign. And if one is prepared to accept the entire set of rules, then Mr Robinson has already done most of the donkey-work in the form of tables showing fuel consumption, gun penetration, the effect of anti-aircraft fire and so on. One would need only to acquire the models at the 400ft to 1in scale recommended, sets of percentage dice and average dice, a tape measure and make up some simple protractor-type devices, to be able to start a complex war game. As with both the Skytrex rules and Mr Hammond's other set, the rules can be adapted for use with models up to 1/1,200 scale.

There is a great deal of detail on some points. Not content with some general advice on simulating weather conditions, Mr Robinson has appropriate effects for shipping for the entire Beaufort wind scale and with probability tables for mist, fog, rain and snow in the four seasons (we are in the north Atlantic). Even the average length of the day for each month of the year is tabulated for us with the corresponding effect this has on visibility and the duration of day-time battles. I would have welcomed more explanation on the application of these weather rules which are otherwise ingenious in conception and set out clearly. With such emphasis on the hazards of the weather, proper attention is given to reconnaissance by aircraft, the use of radar, signalling by radio and night fighting. These and other suggestions, which are wholly for the running of a campaign, are contained in the first quarter of the manual. In the tactical section of the manual there is much detail on firing and its effects. The percentage probability of scoring hits is worked out by ruling lines through base lines for gun bore, for the range and for the target category and thence to a line which produces the percentage score. A point-of-impact chart gives one hundred effects of damage, a one in a hundred score causing the ship's magazine to explode and the immediate destruction of the ship. Damage affects speed, steering, gunnery accuracy, torpedo firing etc, and although these factors are neatly tabulated the effect on the reader is daunting and obviously necessitates a great deal of paperwork for the players. As a general comment on Mr Robinson's rules, I would say that in his desire to offer realistic game rules for every conceivable real-life situation he may have reduced the playability of the game, but he has provided a rich source of ideas for rule-making and brought a more mathematical, and therefore more precise, approach to the whole question of firing which most rules, including my own, lack.

A set of rules for World War II which has won many adherents in Britain is that issued by the Naval War Games Society, Navwar for short. It attempts to be comprehensive so the forty-six page booklet in which the rules are set out is probably best tackled by a new war gamer in small sections. Better still he should endeavour to see the rules in action—they are used both at the Navwar and the 1,200 Model Ship Society meetings. If submarines and aircraft carriers are excluded, a self-contained battle may be played in a matter of a few hours, forty to fifty game moves being possible. But if all the rules are to be employed then only a campaign spread over several sessions gives the requisite time for calculating all the factors involved. The equipment used is modest—a tape measure, a protractor for calculating ships' turns and asdic searches, and a stack of cards for firing effects. When torpedo firing takes place, players also have to place pieces of paper in front of their ships and draw lines for the direction of their shots, having made allowance for the likely position of the target ship when the torpedoes reach its vicinity. Thus torpedo firing does introduce an element of skill into the proceedings. Elsewhere the deck of playing cards play as full a part as do dice in other war games, and for each facet of the game tables are provided by which the draw of the cards is assessed. It is difficult to say what is not provided for by these rules: there are directions for convoys moving in a zigzag fashion, rules for aircraft being launched from submarines—a feature of the Japanese navy—rules for aircraft strafing ships and putting radar out of action. It is apparent, however, that many of the rules work best if an umpire is used, since so much of the game movement is done on maps and involves questions of visibility, reconnaissance and sightings. Curiously the Navwar manual makes little reference to campaign warfare although the rules for calculating fuel consumption, ship repairs and re-ammunitioning are all included and would be used in such campaigns.

The 'Seapower' set of rules for World War II actions was first produced in 1969 by Alnavco, makers of Superior ships, especially for their 1/1,200 scale models. It has undergone various revisions since then and in its current version as 'Seapower III' can be used in battles involving any warships from 1910 to 1945. In some respects the 'Seapower' pack might be called a complete game since all the rules, stationery and equipment necessary for a game are included. But ships are not and there is no doubt that the users of 'Seapower' include many whose fleets are scratch-built or the products of firms other than Alnavco, whilst the amount of space needed to play means that the game can scarcely be regarded as board-type.

All the elements in the game can be bought separately, the most expensive item being the Surface Action and Aircraft Operation Rules for 'Seapower III'. These rule books bring together the basic rules of the game together with suggestions for advanced players.

A second item in the package is the Ship Value Directory. A formula has been applied to hundreds of warships based on the ship's speed, armour and armament and the resulting number is that ship's value for the purposes of allocating ships to each player. Dice are thrown for gun firing, but with the aid of a colouring system for the dice and a Naval Tactics Scoring Sheet results for firing are less a matter of luck than of probability. Factors such as a ship's defensive value and its speed are thus taken into account when deciding how often it has been hit and with what effect.

Movement is by means of theoretical five-minute moves and is measured for each ship on one of three scales according to the space available. A retractable tape measure and a pair of dividers are recommended for use in moving ships, as is a slide-rule for the simple arithmetic computations necessary in the game.

The flexibility of the game means that it can be used by the solo war gamer or by a whole war-games club and in neither case need there be an umpire. A vigorous correspondence amongst users in the Alnavco *Log* over the years has helped iron out anomalies apparent in 'Seapower I', and today the game forms the basis of many war gamers' sets of rules.

Chapter 4

A SUGGESTED SET OF RULES

We have seen how war-game enthusiasts have striven over the years to produce comprehensive sets of rules which make for a fair approximation of naval warfare. Their ideas are certainly worth studying and so are those embodied in the commercial games discussed in Chapter 6. But the reader may find the very comprehensiveness of published sets daunting, and it is for this reason that I offer my own basic set of rules to enable the newcomer to the hobby to get off to a flying start in war gaming. I believe the rules have the merit of being simple to operate, allow players to exercise a fair degree of skill and give a reasonably realistic simulation of naval warfare. Moreover if a small number of ships is chosen for the early actions and modest objectives are set, then the game can be finished in an afternoon or evening session. In time the war gamer will feel dissatisfaction with some aspects of the rules and will make his own amendments, perhaps on the lines of others published. All this is perfectly natural and part of the fun of war gaming, but once a group of war gamers has hit upon a reasonably acceptable set of rules they should adhere to them for a few games at least or the loser in each game will be sure to blame the introduction of a new rule for his defeat!

These rules, then, are best suited to what I believe to be the most common situation in war gaming: the meeting of two or three friends in a modest-sized room for an afternoon or evening battle. A final point about rules—mine or anyone else's. War games can generate heat between the players or end with a note of sourness in the air although not as much, I would claim, as is produced by such a ruthless game as 'Monopoly'! Nor is the provision of a set of rules which allows for every single situation necessarily the answer to the occasional disputes which occur. If the players have a roughly comparable knowledge of naval warfare and are prepared for a little give and take in resolving disputed situations, they will enjoy themselves all the more.

THE PLAYING AREA

I have always used a playing surface marked out in squares and with ships' speeds and weapon ranges therefore given in so many squares per move. I am aware that many other naval war gamers feel that in using this method an opportunity for exercising some additional skill is lost. They prefer to have no set board and make movement and range estimation dependent on the use of the tape measure. The player firing his guns, in fact, calls out his estimate of the target's range, and only after this is the actual distance measured and hits or misses recorded. The player's skill at estimating thus becomes a feature of the game. As against this the players lose the greatest asset of the gridded area which is that submarine warfare and mining operations can be carried out by two players without the use of an umpire since all unseen activities can be plotted on duplicate, and hidden, charts.

For the playing surface I use pieces of thick card of light-blue colour 25in by 20in, each ruled in dark blue to give the same number of squares (eighty) on each. This stiff card was obtained from a high-quality stationers and is tough enough to withstand frequent usage, yet light enough for taking to other war gamers' homes. I can line up two or more sections of card according to whether I want a long and narrow playing area, a square one, a small area or a large one. As they are not joined but lie neatly alongside each other, I can remove areas of sea which are clearly no longer needed as the game proceeds. In order to be able to identify each square for the purposes of submarine searches and the laying or clearing of minefields, I have a series of strips of card in contrasting colours enabling me to identify every square of the board by a code consisting of a letter and a number. The diagram shows this for a playing area consisting of three sections.

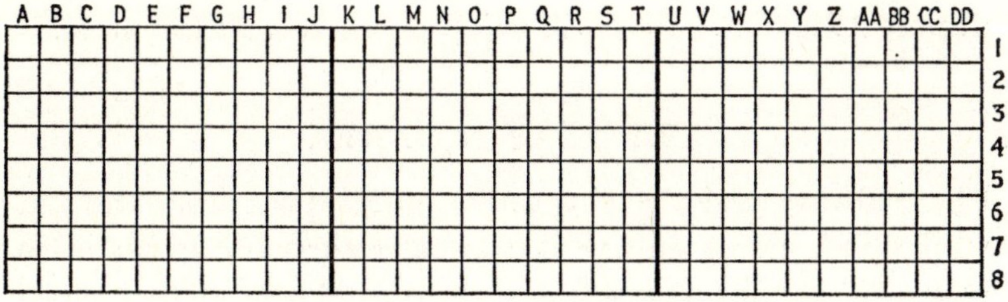

FIG 1 A battle area consisting of three cards

A closer resemblance to the sea can be achieved by using plastic sheeting of a suitable shade of green or blue and painting squares on it or, if it is transparent enough, painting them on another surface underneath the plastic through which they are then seen dimly. Both the cards and the sheeting have the merit of being easy to move and store. I used a chipboard playing surface for some years and found it bulky and almost impossible to use other than in my own home. A last word about the squares. They can be $2\frac{1}{2}$in or 3in square. Any smaller and it becomes difficult to place a 7in long model fairly and

(*Top*) Three popular war-game scales: models of a British County class cruiser in 1/1,200 scale (Eaglewall), 1/3,000 (Navwar) and 1/4,800 (Leicester Micromodels). (*Middle*) Commercially produced war-game ships (1/1,200 and 1/1,250 scales)—top row (from left): battlecruiser *Renown* (Wiking), destroyer (Gillow), aircraft carrier *Enterprise* (Miniships); middle row: cruiser *Biloxi* (Lindberg), battleship *Fuso* (Superior), destroyer *Gravelines* (Triang Minic); bottom row: Narvik class destroyer (Ensign), Town class destroyer (Leicester Micromodels), cruiser *Montecuccoli* (Mercury). (*Bottom*) Scratch-built ships (1/1,200 scale)—top: battleship *Nelson*; centre: aircraft carrier *Eagle*; bottom left: monitor *Roberts*; bottom right: river gunboat *Gnat*. All made by the author

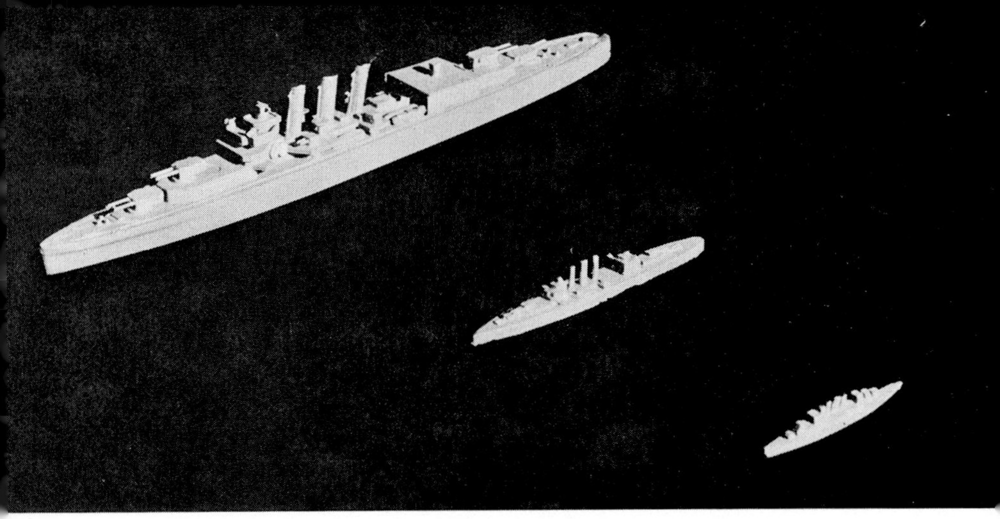

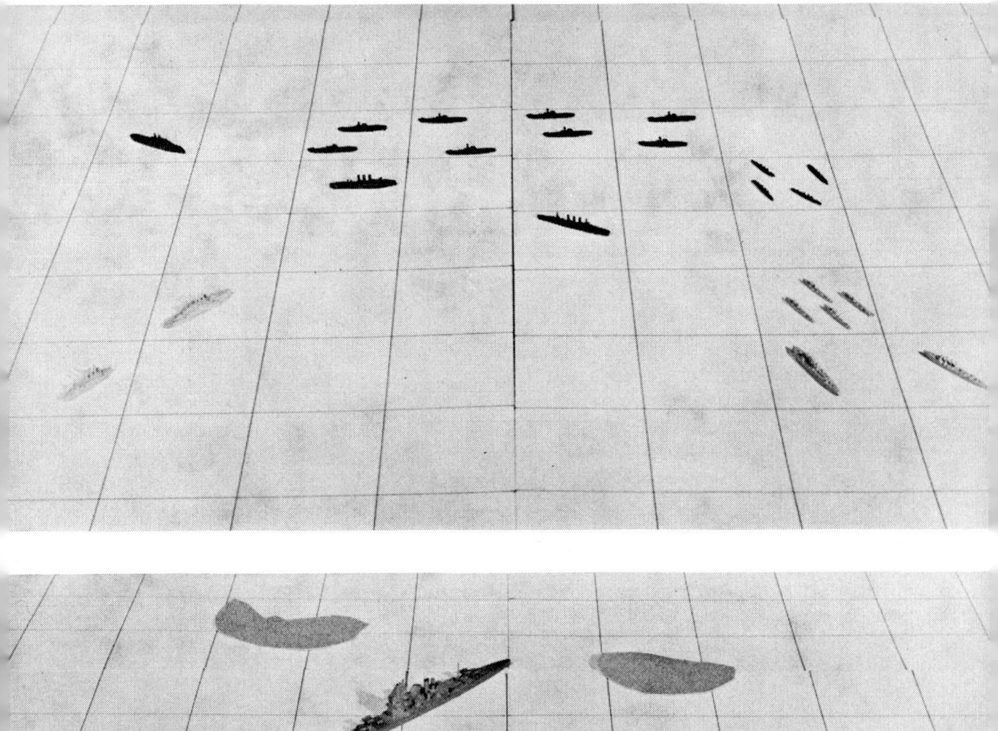

squarely on a square; any larger and the whole playing chart will have to be larger if a satisfactory number of squares are to be used.

HARBOURS, ISLANDS AND COASTLINES In matters of scenic effects, the naval war gamer can rarely equal those of the land gamer. Not for him colourful uniforms, villages, woods, rivers and handy stone walls behind which to shelter troops. But since most naval battles are fought out of sight of all land, the reader need not feel too dismayed.

However, if a harbour is required then strips of balsa wood of appropriate scale to the ships will give a reasonable sea wall. Some chunkier blocks will do for buildings. Both Triang Minic's (1/1,200) and Hansa's sets look much more attractive though with their dock gates, oil pipelines and storage tanks, cranes and nicely detailed breakwater sections. The inclusion of a harbour or two in a game enables ships to go back to them for repairs and for an opponent to have a lovely time laying traps, perhaps with a submarine or a minefield, for the returning ship or for reinforcements hurrying out to sea.

Islands can add interest to a battle area. They may mask a ship's gunfire, or a player may skilfully deploy mines and ships between them in order to lure his opponent into a battle zone of his own choosing. The capture of a certain number of islands may be the objective in a particular game. I keep to hand a supply of islands cut from card which fit approximately any combination of squares. A certain amount of artificial shaping

(*Top*) Professionally made wooden ships (1/1,200 scale)—top: battleship *Richelieu*; left: cruiser *Oakland*; right: destroyer *Radford* (all British Admiralty recognition models). (*Centre*) Adriatic battle scene (1915)—the Austro-German convoy is under attack from Italian warships (bottom left) and British warships (bottom right)—all ships by Skytrex (1/3,000 scale). (*Bottom*) Nicobar Islands action (1942)—the raider, *Atago* (left), fires her ten 8in guns at the cruisers *Achilles* and *Leander* before veering away (*Atago* by Superior, others by Eaglewall)

ensures that one square can be used as sea whilst another is deemed to be wholly land. Islands can also be made three-dimensionally using layers of contoured polystyrene, coloured for a more attractive effect.

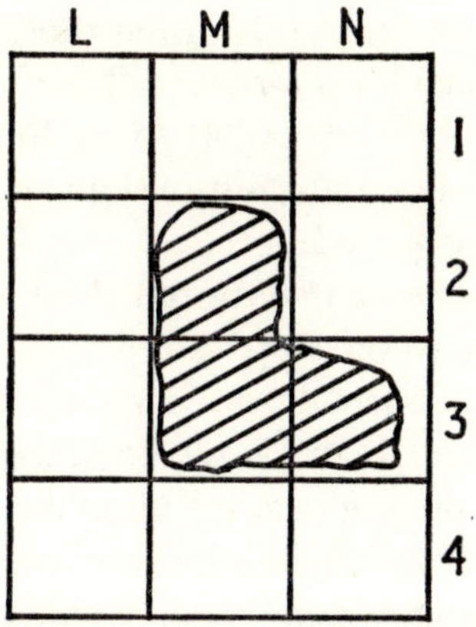

FIG 2 The island covers M2, M3 and N3

Lastly, coastlines. Larger pieces of card can be used to provide these, so can a piece of string sufficiently heavy to lie flat. On certain war-game surfaces, for example, a wooden or stone floor, a chalk line will do admirably and can be rubbed out after the game.

OBJECTIVE OF THE GAME

This is generally agreed upon in advance, but if an umpire is being used then he can be asked to define reasonable conditions for victory for each side and announce when one or the other has been attained. Objectives can be as simple as sinking more of the enemy's ships than one loses, getting a certain proportion of a convoy from one place to another, or hunting down a raider. Or points can be given for attaining a number of objectives, victory going to the player first attaining the agreed total quota. I have always admired Waddington's game

'Buccaneer' which, set in the sailing-ship era, allows each player to choose any combination of methods to attain the winning points total of treasure. By building into a game more than one possible objective, players find there is more scope for independent action. Conversely, if the victory condition is such that it encourages a massive confrontation of all the ships taking part in the game, then the action will be tedious indeed.

ALLOCATING THE SHIPS

Many war gamers find it simple enough to share out ships before a game in an equable fashion merely by knowing the capabilities of the real ones. For example, one British battlecruiser of World War I can be equated with a German one with smaller-calibre guns because the latter was better protected. Or one *Bismarck* may be said to equal one *Hood* and one *Prince of Wales*, but only if the rules allow for *Hood*'s vulnerability to plunging fire and the *Prince of Wales*'s rawness. In any case, it is neither essential nor necessarily desirable to have equal sides in all game situations. But the reader who prefers to apportion ships according to more precise values can draw up a formula which gives each ship a points value based on its capabilities. It might follow the lines of Fletcher Pratt's formula, mentioned in Chapter 3, or it can be made somewhat simpler. Such a formula requires that each characteristic of the ship be given a points value, and that these points be aggregated to give a grand total. Thus such factors as a ship's speed, its resistance to damage (sinking rate), the guns, torpedoes, aircraft and mines it carries, and its anti-aircraft rating will all be calculated. But I must confess that I have never been entirely happy with the end results of my own formula and still prefer to apportion fleets after amicable bargaining with an opponent.

SHIP CARDS

At this point in the preparations for a game, may I recommend that the war gamer prepare a card for each of his ships? Sets of cards can be bought at any stationers and can be stored in a little box-file or shoebox for neatness. Each card should show the ship's name, displacement, weapons, speed, distribution of armour, and any important variants amongst ships of the same class. Then, in pencil, the gamer can add details relevant to war gaming, for example, the points needed to sink a ship, the anti-aircraft rating of the ship, the points value of the ship for allocation purposes. I say in pencil because the war gamer is

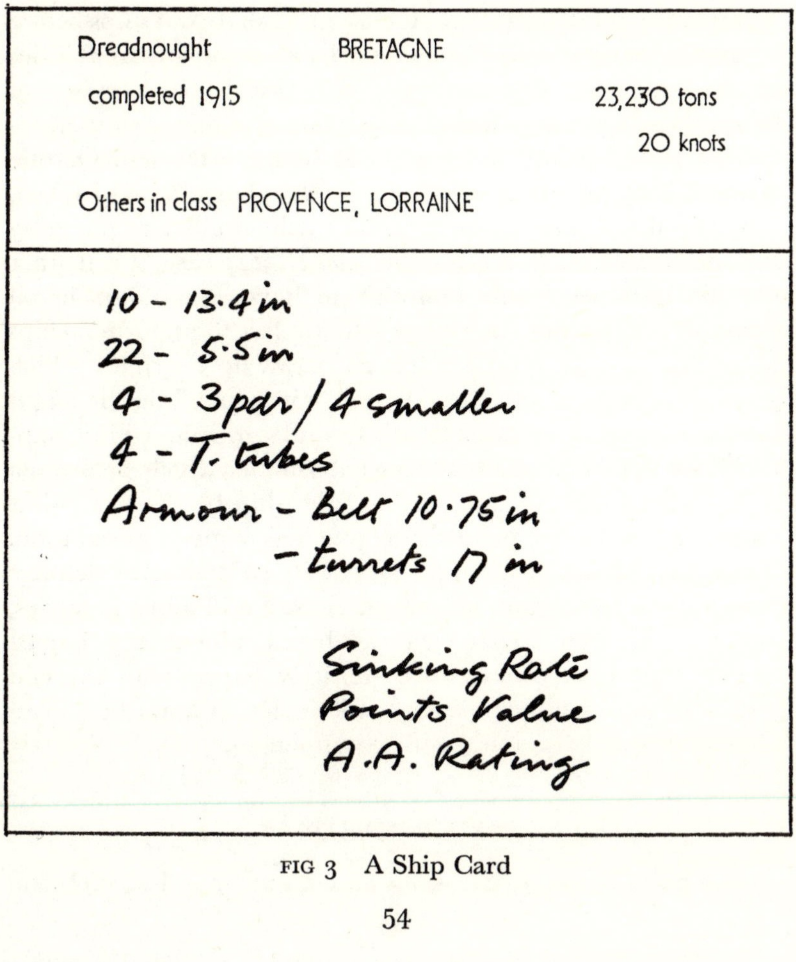

FIG 3 A Ship Card

almost bound to have second thoughts about these various values as he gains experience and amends his rules, and will want to alter that part of the card. The use of cards makes it easy to hand a player the complete details of his fleet for a particular action, whereas if these are contained in your master set of rules or war-game file you will waste time consulting them from amongst many others. The card looks like the one in Fig 3.

For each game it is necessary to record damage to ships, the numbers of torpedoes used up, adjustments to speed. I use simple sheets which I have had duplicated and on these the combatant ships are entered. Fig 4 shows how an American squadron's sheet looked at the end of a successful defence of a convoy against a Japanese attack. This shows that *Iowa* sank after sustaining more than 900 points of damage, most of it coming from six torpedo hits of 120 points each. *Oakland* was sunk by the battleship *Musashi*, apparently without having fired her torpedoes, whereas *Jenkins*, sunk by assorted gunfire, had loosed off both sets of her ten torpedo tubes. *Fletcher* had

Name	Turrets knocked-out	Torpedo tubes	Damage sustained
~~Iowa (900)~~			720 points from tubes
Yorktown (150)			
Enterprise (150)			
Little Rock (110)			
~~Oakland (80)~~		~~6~~ ~~6~~	Sunk by Musashi (13¾ points; 1 square in speed)
Fletcher (30)		~~5~~ ~~5~~	
Radford (30)		~~5~~ 5	
~~Jenkins (30)~~		~~5~~ ~~5~~	8¾ 29½ 33¾
LaValette (30)		5 5	
Chevalier (30)		5 5	

FIG 4 A Damage Sheet

used up her torpedoes and had suffered enough damage to lose one square in speed per game move. *Radford* had fired five of her ten torpedoes.

We are now armed with a playing surface, ships' record cards and sheets for recording damage in combat. A dozen dice, preferably of the large variety, and a cup which can take them all if necessary, completes the equipment for the moment.

Your ships are in harbour, eager to set off for what may well prove to be costly combat. What happens now?

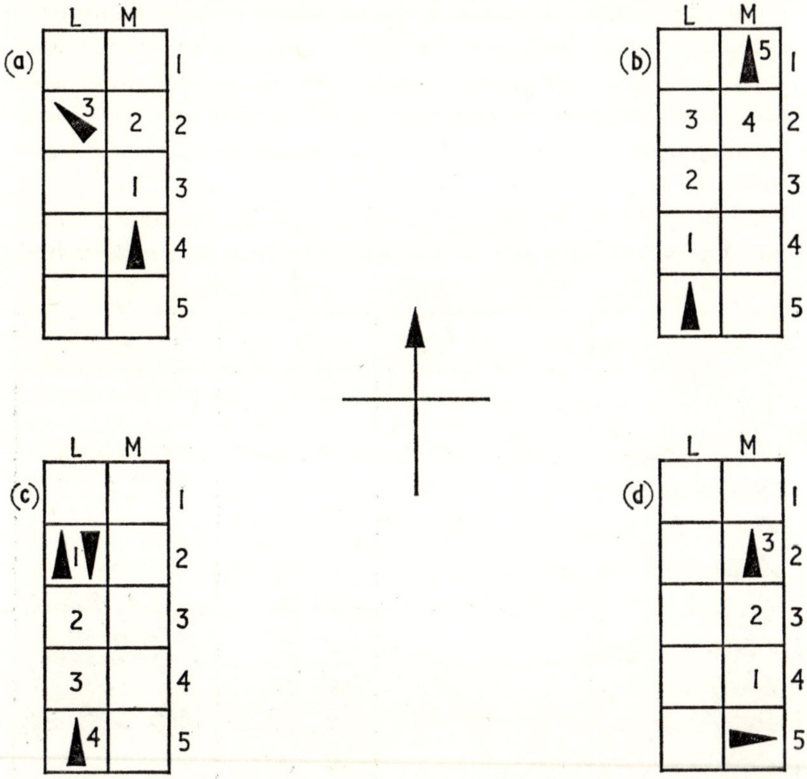

FIG 5 (a) Ship moving 3 squares to north-west; (b) ship moving 5 squares to north-east; (c) ship 'turning tail' and moving 4 squares; (d) ship moving 3 squares to north

MOVEMENT OF SHIPS

Each player moves any or all of his ships up to the maximum number of squares allowed. Changing course is rather like making the knight's move in chess, and if a 180° turn is involved one square in speed is lost. Some examples are shown in Fig 5. The ship should be left unambiguously at right angles to the sides of the square or on the diagonal as this affects its vulnerability to torpedo attacks and its movement next turn. The ship's bridge and funnels should lie in the centre of the square. Except in harbour or when re-provisioning at sea, no ship of above 3,500 tons can occupy a square with another. In practice this means two destroyers, four motor torpedo boats or one cruiser or larger occupy a square. Ships can, however, pass through a square occupied by other ships.

So our fleet moves out to sea. Let us say that it consists of a battleship, two 6in gun cruisers and four destroyers. It might be disposed as in Fig 6 to give reasonable protection to the battleship. The destroyers will be using detection equipment to ask if a submarine is under any of the squares over which they pass although such equipment would not be available for World War I situations. The cruisers are protecting the battle-

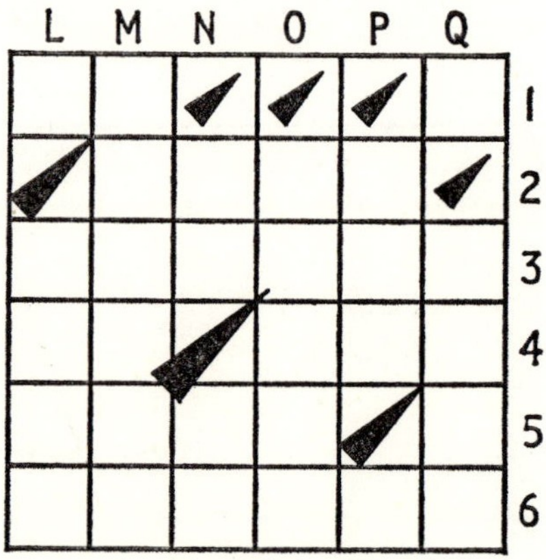

FIG 6 Fleet moving out to sea

ship's flanks from destroyer attack and can be sent off quickly to investigate anything suspicious. The battleship is a 25 knot vessel, the cruisers 32 knot vessels and the destroyers 35 knot vessels, so if the group is to move along together it will travel at the battleship's speed. The normal range of speeds encountered in the game can be summarised in a table:

Actual speed in knots	*Speed in squares per move*
Up to 7	1
8–11	1/2
12–15	2
16–19	2/3
20–23	3
24–27	3/4
28–31	4
32–35	4/5
36–40	5
41 and above	5/6

To employ this wide range of speeds I give myself the slight inconvenience of having to move some vessels more squares in one move than in the next. As an *aide-mémoire* a card is displayed as in Fig 7 and it is turned round on the conclusion of each game move. Players could take it in turn to move first or dice for choice each time. I believe that a good compromise is to go first

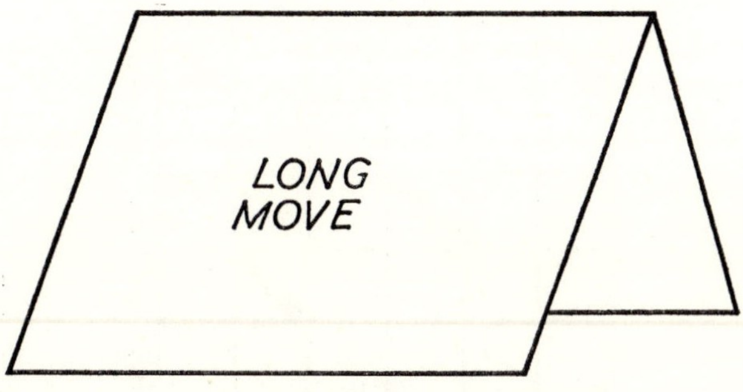

FIG 7 Long-move card

alternately. The sequence of moves, together with any fighting necessary, becomes

Move 1 (*long move*)	Player A moves Player B moves Combat resolved
Move 2 (*short move*)	Player B moves Player A moves Combat resolved
Move 3 (*long move*)	Player A moves Player B moves Combat resolved

SINKING THE ENEMY'S SHIP

All ships have a sinking rate, a number which once exceeded in damage points whether from shells, torpedoes, bombs, depth charges or mines means that that ship is sunk. Even the small guns of destroyers are deemed capable of causing larger armoured ships damage perhaps in damaging equipment, killing personnel, causing fires, all of which can contribute to a ship's destruction. As progressive stages of the total are passed so turrets are deemed out of action and the ship's speed reduced. How the reader arrives at a sinking rate for each ship is a matter of personal preference. I like my battleships to have a rate in the range 700 to 1,200 points. Smaller vessels are then scaled down accordingly. These figures were arrived at originally by estimating how many points of torpedo hits or gunfire might realistically be regarded as sufficient to sink different types of ship. Thus if your reading suggests that a World War II battleship of average proportions and armour would sink after nine torpedo hits, then its sinking rate can be regarded as nine times one conventional torpedo's damage capability, ie 9 × 100 = 900 points. A destroyer of 1,300 tons or so has a sinking rate of fifty-five points.

Other war gamers attempt to arrive at a more accurate assessment of a ship's capacity to absorb punishment by considering all its known armour characteristics and relating these by a formula, much as Fletcher Pratt did in his rules. In the table below I give some specimen sinking rates for some warships in the period 1914 to 1945. They were arrived at by consulting reference books for the armour distribution of each vessel, the degree of compartmentalisation and the actual performance of a ship under fire if known. The last point is an interesting one. Some warships served their turn and passed on into history without ever having come under severe attack. Others, such as the German battlecruiser *Seydlitz* in World War I took a tremendous pounding and testified to the excellent design features of this class of ship.

SUGGESTED SINKING RATES

World War I	
Audacious	500
Kaiser	750
Nevada	750
Warspite	640
Yamashiro	650
Indefatigable	350
Invincible	320
Seydlitz	500
Minotaur	200
Scharnhorst	220
Arethusa	100
Bremen	120
destroyers	40
submarines	10

World War II	
Fuso	550
Jean Bart	800
King George V	750
North Carolina	900
Roma	750
Royal Sovereign	600
Scharnhorst	700
Yamato	1,200
Essex	250
Hermes	150
Illustrious	250
Sangamon	150
Shinano	500
Zuikaku	250
Atago	200
Bolzano	180
Dorsetshire	200
Oakland	130
Prinz Eugen	280
Asashio	60
Fletcher	60
Hardy	55
von Roeder	60
corvettes	50
trawlers	25
submarines	10

FIRING GUNS

Let us say that our squadron of ships has come into range of an enemy warship—an 8in gun cruiser. Combat occurs once both players have completed their moves. It does not matter who elects to do the firing first as the firing is deemed simultaneous so that no ship is removed from the board as sunk or loses turrets until all firing on both sides has taken place. Ranges are soon memorised. The following are based on official data wherever possible, but some sources vary in estimates of how far guns could fire. The examples are for the most common naval guns of both wars and are based on the premise that 5,000yd actual range is equal to one game-board square.

World War I		World War II	
Calibre in inches	*Range in squares*	*Calibre in inches*	*Range in squares*
18	9	18	9
15/14	5	16	9 *Iowa* 7 *Nelson*
13·5	4		
13 down to 8	3	15/14	7
7·5 down to 4	2	15	5 R class battleships
3 and below	1	11	7
		11	3 *Schliesen*
		8 down to 5·1	5
		5	4
		4·7	3
		4	2
		3	1

From the above list it will be seen that some guns of seemingly identical calibre will vary in range according to their age or elevation allowed by the turrets. The British Queen Elizabeth and Royal Sovereign classes had the same 15in guns in both wars, but the turrets of the former had been modernised so the guns' range far exceeded those of the unmodified class of battleships.

A vessel in range of another fires any of the main armament which can be brought to bear and a dice is thrown for each gun fired. Ranges are counted out in squares as with movement (see Fig 8 below). If the guns are firing at maximum range

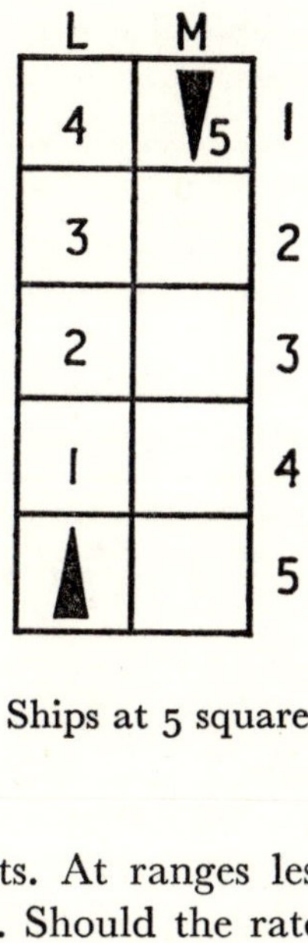

FIG 8 Ships at 5 squares range

only sixes count as hits. At ranges less than the maximum, fives also count as hits. Should the rate of fire of a particular calibre of gun be such as to allow a further third or fourth set of throws, then another target can be engaged applying the same criteria as above. But if the second, third or fourth firings are aimed at the same target, then the accuracy of their fire is deemed greater and fours, fives and sixes all count as hits. The following table shows what damage is done by a shell of a particular size. The figures are based approximately on the known weights of shells. The rate of fire of various guns is given too.

World War I

Gun calibre in inches	*Rate of fire per move*	*Damage points per hit*
18	1	64
15	1	39
14	1	28
13·5	1	28
13	1	23
12·5	1	20
12	1	17
11	2	14
10	2	10
9·4/9	2	8
8·2	2	6
8	2	5
7·5	2	4
7	2	3
6·7	2	2½
6·4/5·9	3	2
5·5	3	1½
5	3	1
4·7	3	⅔
4·1	4	½
4	4	½
3·5	4	⅓
3	4	¼

World War II

Gun calibre in inches	*Rate of fire per move*	*Damage in points per hit*
18	1	64
16	1	54 *Iowa* 49 *Rodney* 45 *Nagato*
15	1	39
14	1	32
13·5	1	25
13	1	25
11	2	13
8	2	5
7·5	2	4
6·1/5·9	3	2
5·5/5·25	3	1½
5	3	1
4·7/4·5	3	1
4/3·9	4	½
3	4	⅓

It will be readily seen that the information contained in the two previous tables can all be placed in columns on one large piece of paper or card and can thus be consulted easily during a game. After the main armament has been fired the same procedure is followed for any secondary or tertiary guns which can be brought to bear.

FIRING TORPEDOES

Torpedoes may be fired when the range permits. Their range and capability for damage is:

Torpedoes	*Range in squares*	*Damage in points*
All World War I	2	70
All World War II (except Japanese)	2	100
Japanese of World War II	4	120

I insist that torpedoes fired from ships are fired as sets. Thus, a destroyer with two sets of four tubes must fire all four of one set at the same target and only then switch targets and fire the other set. Or, of course, he may fire both sets at the same target. Used in this way some torpedoes will be wasted as one hit alone might have sunk the target whereas two or more torpedo hits actually registered. But this makes for interesting decisions often taken by the player concerned at great speed as he strives to inflict the most damage to the most ships for the least expenditure of his fire power. In contrast to gunfire I make allowance for the angle the target presents to the attacking ship when deciding the chances of a torpedo hitting it. It is important therefore that all ships are placed quite unambiguously on their squares at the end of the movement turn. The three examples of a ship attacked by torpedoes in Fig 9 apply whether the torpedoes are launched by surface vessel, submarine or aircraft.

While all this firing goes on the players must keep a careful account on their damage sheets of damage incurred and torpedo tubes fired. Gun turrets are knocked out, starting with the

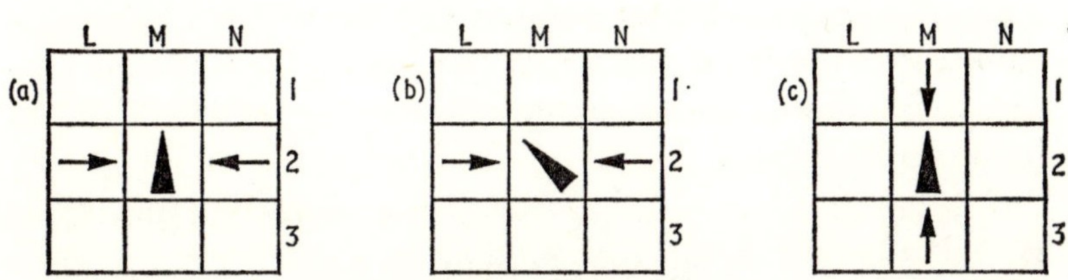

FIG 9 (a) Ship attacked on either beam. Two dice thrown for each torpedo, 9, 10, 11, 12 count as hits; (b) ship attacked from oblique angle, 10, 11, 12 to hit; (c) ship attacked from ahead or astern, 11 and 12 to hit

main ones of the defender's choice, and speed is reduced at the following rates.

EFFECTS OF DAMAGE ON SHIPS

Percentage of sinking rate	*Reduction of turrets/speed*
20	— 1 turret
40	— 1 turret/one square off speed
60	— 1 turret
80	— 1 turret/further square off speed
90	— 1 turret
95	— 1 turret/cannot fire tubes, speed to one square only

Note: a ship of very slow initial speed which is reduced to one square after only receiving 40 per cent of damage becomes totally immobile after 80 per cent damage.

When the firing is completed, sunken ships are removed. It is a matter of choice whether players mark any ship with damage by some kind of symbol, say cotton-wool for smoke, or add to the fun of the game by ruling that a player may only learn of a ship's condition after he has started to fire at it in the next combat turn.

TAKING ON MORE TORPEDOES

I only allow a torpedo-carrying vessel, except the submarine, to have torpedoes sufficient for one firing from each of its tubes. If it needs more it must return to port and stay there for one movement turn or go alongside a depot ship at sea. In the latter case it may move at the depot ship's speed but cannot use its armaments in the game move during which it takes on extra torpedoes. Surfaced submarines may also use a depot ship's facilities but may not submerge until the loading operation is completed.

SUBMARINES AND SUBMARINE HUNTING

Excepting early World War I and late World War II types I have submarines moving at two squares when surfaced and one when submerged. Another simplification is to allow all submarines three sets of four tubes, twelve torpedoes in all. Because of these low speeds the player using submarines has to be patient and sometimes will be more successful if he lures enemy ships over them by baiting the trap with a tempting surface target, rather than in trying to chase fast-moving ships. If submarines are always obliged to commence operations from their base port they will rarely get into the action so it is advisable, and adds to the general uncertainty of the game, to allow players to place two or three submarines submerged in any part of the playing area at the start of the game. Underwater movement can be recorded by using a smaller replica of the main game board and moving discs on this instead of models. This replica chart is, of course, kept carefully out of sight of the opposing player. Since both the game board and its replica have a letter and number code for each square, it is easy for the hunting player to ask if his ship is on a particular square and for the submarine commander to answer yes or no. It is a useful device for the submarine commander to affect the same unemotional voice when answering each search request lest his tremulous voice give a clue to the hunting player. Needless to say, the frenzied lighting of a cigarette may equally help the searcher.

(*Top*) Harbour scene—the battleship *Warspite* is manoeuvred seawards by four tugs. The destroyer *Electra* takes on oil, the cruiser *Exeter* is in dock (warships by Eaglewall; tugs by Triang Minic and Ensign; harbour accessories mostly by Triang Minic). (*Centre*) Atlantic rendezvous—the German pocket battleship *Admiral Scheer* closes in on the depot ship *Dittmarschen* to take on supplies (both ships by Eaglewall). (*Bottom*) The search board (right) and battle board used for this World War II game, 'Midway', by the American firm of Avalon Hill

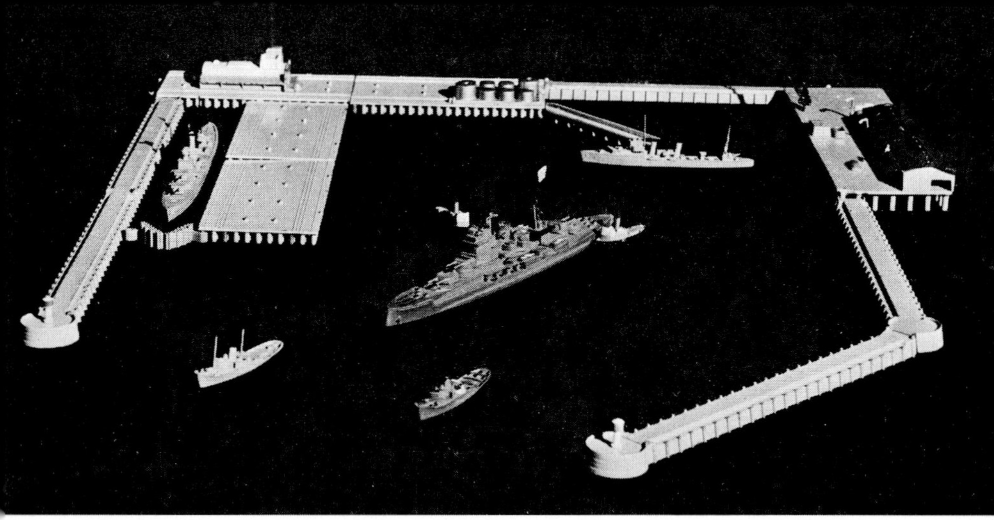

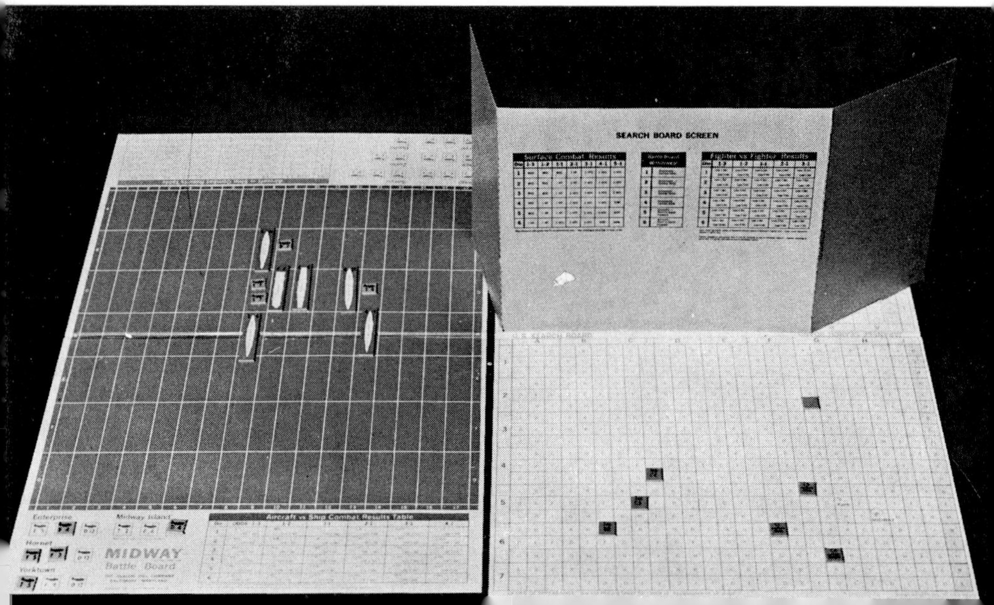
SEARCH BOARD SCREEN
Surface Combat Results
Fighter vs Fighter Results
Enterprise
Midway Island
Hornet
Yorktown
MIDWAY
Battle Board
Aircraft vs Ship Combat Results Table

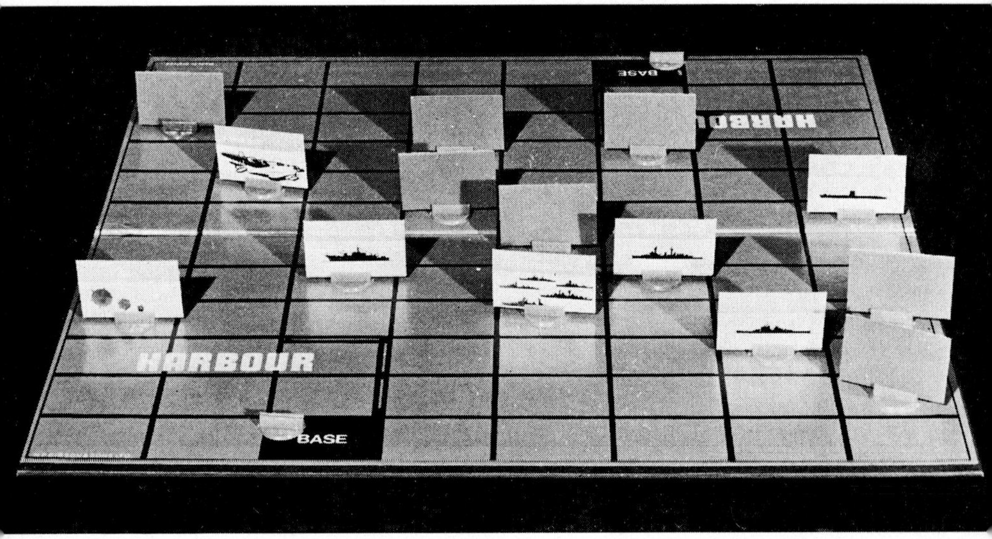
HARBOUR
BASE

Once a submarine has been located, that is an appropriately equipped ship has elicited the answer yes when enquiring of the square it stands on or is passing over, attack and evade procedures begin. The surface ship may fire depth charges in the combat move, one dice being thrown for the attack. A score of six sinks the submarine, a five forces it to the surface with four points recorded as damage, and it may only submerge after missing a movement turn and not at all if its total damage reaches 60 per cent. But should the first depth charge attack fail then the submarine may yet escape. Even at a speed of one square the submarine can be on any one of five squares in the next move, and the search procedure will have to begin again. If there is only one ship searching, the submarine can often get clean away or even sink the pursuer. But the determined submarine hunter will only go searching with two or more vessels and will thus be able to corner the submarine very quickly. Fig 10 illustrates this sequence of events.

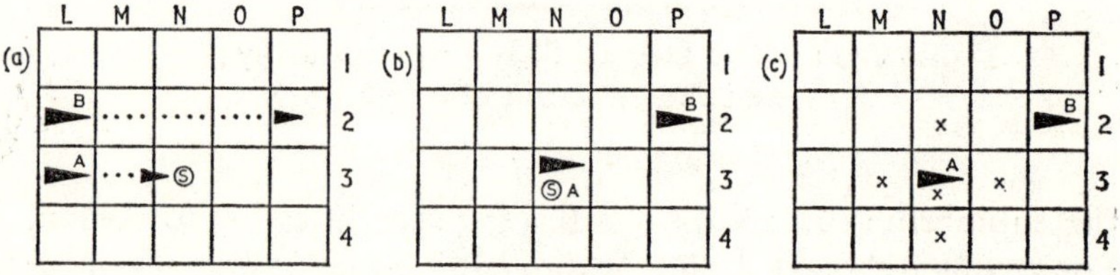

FIG 10 (a) Destroyers searching A get a yes; (b) A attacks—B has exhausted its movement; (c) next turn—submarine will be on one of 5 squares (x)

(*Top*) U-boat attack!—a freighter lies dead in the water whilst the British frigates *Magpie* and *Kite* turn to chase the German U-boat which is still on the surface (warships by Eaglewall; freighter—'Hong Kong miscellaneous'). (*Centre*) The closing stages of the 'Dover Patrol' board game by the British firm of Gibson. (*Bottom*) The *Scharnhorst* is at sea!—the German battleship *Scharnhorst* is escorted by four destroyers and the cruiser *Leipzig* (battleship by Superior; destroyers by Eaglewall; cruiser—scratch-built)

The submarine attacks with torpedoes in the same fashion as surface vessels—the target can be on the same square when twelve to nine scores all count as hits, or one or two squares away. As with surface firing, the range is calculated without using diagonals (see Fig 11). After a set of tubes has been used,

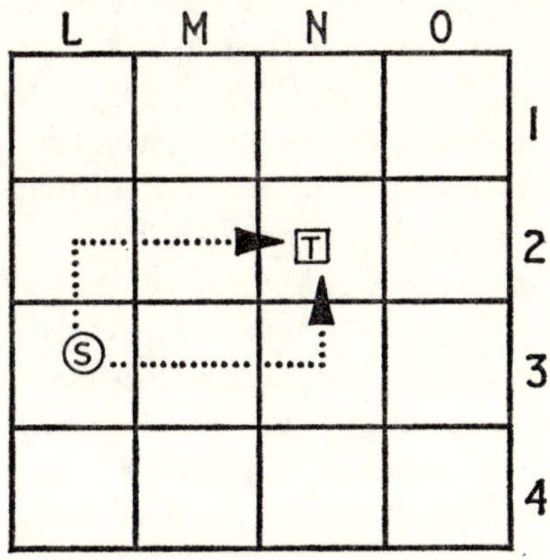

FIG 11 The target (T) is 3 squares away and therefore out of range

the submarine must miss one firing move in order to reload. For simplicity's sake I allow submarines to stay submerged for indefinite periods and for attacking vessels to have unlimited numbers of depth charges. In a game wholly devoted to submarine-versus-hunter tactics, I put limits on both these factors and also allow submarines to work at two other depths. When at the second depth they can be detected, but can neither fire nor be hit by depth charges. At the third depth they must remain stationary, cannot fire and cannot be detected. To change depths takes one movement turn. The determined hunting vessel will therefore have to stay in the vicinity until his quarry comes up for air, but in doing so it may become the victim of another submarine.

USING AIRCRAFT CARRIERS

I only use aircraft carriers sparingly as the great range advantage planes have over guns can produce some dull actions in which conventional surface warships never get to grips with each other. In fact I deliberately scale down aircraft ranges so that carriers have to be put well up in the action area if they are to be able to recover their aircraft. They then become vulnerable to surface attack and their commander has to make some rather precise calculations for the aircrafts' sake and for the ship's.

Planes may take off from the carrier at any stage of the movement turn, their presence in flight being marked by a numbered or lettered disc on the playing area. On the reverse side of the disc the player with the aircraft must have noted the combination of aircraft in the group (Fig 12)—numbers are revealed to his opponent when combat occurs. I allow all carrier aircraft to be used in any of the three roles of fighter,

FIG 12 Air strike

dive-bomber or torpedo carrier although once loaded up and in the air the roles cannot of course be switched. Fighters are entitled to flights totalling forty squares, consisting of four game moves of ten squares. The more heavily laden dive-bombers and torpedo planes can travel thirty-six squares in four game moves of nine squares each. Carriers launch and recover planes without alterations of speed or direction—another distortion of real practice for the sake of simplicity, but aircraft cannot take off again until one movement turn has passed during which they are refuelling and rearming. If a carrier suffers damage its flight facilities and speed are affected as follows:

Damage (percentage)	
40	one square in speed lost; flight facilities lost for one game move
60	flight facilities lost for two moves
80	speed reduced by a further square; flight facilities lost for three moves
90	speed reduced to one square; no further flights possible

Aircraft move in the same manner as ships, that is to say, in straight lines and not through diagonals. If an aircraft still has some movement squares unused after making its attack, it may move off to the appropriate limit.

Bombs and torpedoes are launched when the aircraft is on the same square as the target vessel. But whereas the torpedo-carrying aircraft operates under the same rules as when surface vessels make attacks with torpedoes, dive-bombers encounter special difficulties according to the speed of the target. The table shows this:

Speed of target (knots)	*Score required for hit*
32 +	one die, 6 scores
20 +	one die, 6 and 5 score
8 +	one die, 6, 5 and 4 score

Aerial torpedoes and bombs inflict the following amount of damage:

Torpedoes		*Bombs*	
Japanese	120 points	500lb	35 points
Others	100 points	880lb	62 points
		1,000lb	68 points

DEFENCE AGAINST AIRCRAFT ATTACKS

The attacked ship will, of course, endeavour to fight off the aircraft, and here anti-aircraft ratings, which were shown on the Ship Card in Fig 3, come into prominence whilst ships may also manoeuvre to give themselves mutual protection.

Anti-aircraft ratings are given to each vessel and put on their Ship Card according to the known numbers of anti-aircraft guns carried by the real ship. As a rough rule of thumb I give one point for ten anti-aircraft guns carried, giving extra credit for the high calibre of some or the superlative qualities of others, such as the 40mm Bofors/Hazemeyer guns of World War II. In no case have I given more than fifteen as an anti-aircraft rating (to the *Yamato* and *Musashi* battleships). *North Carolina* gets twelve, *Warspite* only four. Cruisers with dual-purpose armament and sophisticated control systems, such as the American Oakland class and the British Dido class rate higher than many old battleships. To fire at an attacking aircraft, the ship concerned and the ships immediately adjacent to it and in the attacker's path may throw dice at the rate of one for each point of their anti-aircraft rating. And because torpedo planes are generally in sight longer than dive-bombers, the defenders may have two bursts of fire against them or one against them and one against dive-bombers. Aircraft are destroyed as follows:

Scores required	*Aircraft destroyed*
6, 5 and 4	torpedo plane
6 and 5	dive-bomber
6	fighter

A fighter is only likely to be in range of ship fire if it has been engaged in combat over the ships concerned. Additional opposition for the aircraft is provided by the rule already mentioned above that any ship or ships adjacent to the target and in the attacking plane's path may also fire. Fig 13 shows how some ships can be deployed to give maximum protection to the battleship. We have the same group of ships mentioned at the beginning of this chapter, and from whichever direction an air attack comes the defending ships can put up anti-aircraft ratings totalling eleven or twelve points.

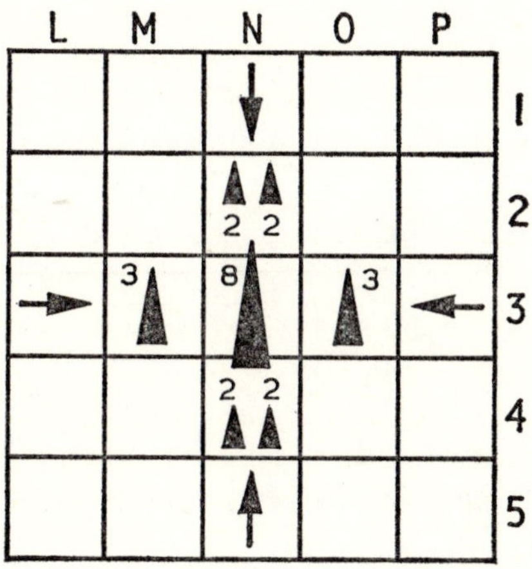

FIG 13 Anti-aircraft protection for battleship on N3

How can the aircraft counter such defensive arrangements? One solution is to throw every available aircraft into attack and all from the same direction. Sheer weight of numbers may then overwhelm the ships' defences. But I regard any group of planes exceeding forty as likely to get in each other's way so that for every aircraft over forty making a simultaneous attack a dice is thrown. A six means two planes collide and are destroyed, a five or a four that the area was so congested that the plane could not launch its bomb or torpedo. A more subtle tactic, in any case, would be to attack the screening vessel or vessels with some aircraft and only then to attack the battleship with the remainder. But tactics for both attacker and attacked evolve during successive war games and are part of the enjoyable process whereby one game's successful ploy may prove to be the next one's disaster.

AERIAL COMBAT

Devotees of the Pacific War of 1941–5 will be asking, what of combat between carrier aircraft? There have been many war games in which to make for faster, less complicated play I have had no element of air-to-air combat. Carrier planes have made their attacks against ships, suffered casualties and returned to their floating base. But in real warfare, aircraft may never even reach their target if the other side's 'combat air patrols' are active, and I now use the following simple rules to resolve aerial combat.

Combat occurs when, at the end of a movement turn, aircraft of opposing sides are on the same square. If the groups consist of several types of aircraft then fighter-versus-fighter combat is resolved first, and then in the second part of the combat phase the surviving fighters may elect to attack dive-bombers or torpedo planes. When any two planes fight each other, each player throws a dice and kills are scored as in the following table. Naturally fighters are given better odds when fighting more vulnerable planes.

Fighter	against fighter, 6 and 5 destroy
	against dive-bomber, 6, 5 and 4 destroy
	against torpedo plane, 6, 5, 4 and 3 destroy
Dive-bomber	against fighter, 6 destroys
	against dive-bomber and torpedo plane, 6 and 5 destroy
Torpedo plane	against fighter, 6 destroys
	against dive-bomber and torpedo plane, 6 and 5 destroy

In an action where, say, eleven planes attack eight, plane-to-plane combat is resolved for eight planes on each side and the spare three then engage any survivors in a second round of combat. Alternatively, the spare ones are used in one or two particular dogfights so that the player concerned can make two or more attacking throws of the dice to his opponent's one.

Aerial combat should precede attacks on ships and the use of anti-aircraft defences, but the attack on a ship can come all in the same combat move as the aerial battle. Players may keep a regular patrol of fighters in the air over any of their own ships up to the maximum of four game moves or go out and intercept attacking aircraft.

I leave carrier and aircraft rules with a word of warning. The comparatively simple rules I have outlined all take up a fair amount of time to implement. It sometimes makes for a better game if only one carrier, or even a slow escort carrier, is assigned to a player so that its use is more carefully thought out and game moves are not dominated by air battles.

I conclude this chapter with some minor rules although these will not account for every facet of naval warfare and for which the reader will no doubt draw up his own.

REPAIRS

To regain the ship's full sinking rate the damaged vessel must return to port and miss one complete game move for each 10 per cent of the total sinking rate restored. Its weapons, except anti-aircraft ones, cannot be used in harbour. It may leave harbour at any stage of its repairs and will sometimes need to do so if a desperate situation has arisen at sea.

MINELAYING AND MINESWEEPING

A square of sea is deemed sown with ten mines if the minesweeper of thirty knots or more passes over that square. If a slower vessel is used, then it must stay on the square for one whole move. The ship's cargo of mines is used up at the rate of ten mines per square sown.

When an enemy ship passes over a mined square, the player announces the fact. Two dice are thrown. All scores except two sixes are deemed hits, the aggregate being multiplied by ten to give the damage registered. Thus the least a ship will be damaged is 10 (1 + 1) or 20 points, the most 10 (6 + 5) or 110 points. The former was probably a mine detonating at some distance, the latter a direct explosion from underneath. The minefield remains active until ten vessels have passed over it or it is cleared.

Only vessels designated as minesweepers can go on a mined square and not suffer damage. It takes one minesweeper two whole game moves to clear a square, but as minesweepers are generally small vessels, up to four could be used on any square and thus it could be swept in just the time the ships take to pass through the square.

CHANGING WEATHER CONDITIONS

These are easily introduced although they are not always worth the bother in a short game. The playing area is divided into four zones for weather purposes. Before each game move a dice is thrown for each zone with the following results: six denotes fog; five and four denote high seas; three, two and one denote good visibility and moderate sea conditions.

For one whole game move the appropriate weather prevails in each of the four zones and cards showing this are placed alongside the zones. Where high seas are running, all ships below 10,000 tons displacement drop a square in speed and carrier aircraft may not be launched or retrieved whilst submarines cannot fire torpedoes. Where fog is the hazard all ships reduce their speed to one square per move, all planes are confined to their carrier, no ships can fire except for their depth charges and ships are removed from the playing area and replaced by discs for that move. This introduces an element of hidden movement especially if the next game move brings a continuance of the fog or some discs/ships can move into a zone where fog prevails.

NIGHT ACTIONS

Again, these are best used when time is not at a premium. Four whole game moves make up one night, eight one whole day. Movement by night on the playing chart is with discs, not models. The name of the ship represented by the disc is written on the reverse side and is, of course, known only to the player of that fleet. Planes are grounded.

HIDDEN MOVEMENT TO START A GAME

A screen placed across the centre of the playing chart for the first two or three game moves allows players to deploy their ships secretly and thus makes subsequent moves more interesting. A variation on this is to dice for the point at which a ship or ships will enter the playing area. This will sometimes put ships closer to each other than the competing players would have preferred.

Chapter 5

SOME ACTIONS USING THE AUTHOR'S RULES

In the two games I describe in this chapter the various rules introduced in Chapter 4 are amplified. At the same time I try to convey something of the flavour of keenly fought battles. Each battle illustrates a different type of game although in each a definite objective was set so that the fighting would not justify a general mêlée in the centre of the playing area. The situations described are imaginary, but all the ships engaged are actual warships of the periods concerned and the scenarios are inspired by authentic events.

The chapter concludes with suggestions for two quite different sets of circumstances—for war gamers who cannot always find opponents there are some hints on solo play, and for those who want to mount a whole campaign an example of the sort of planning that is necessary is given.

GAME 1: THE DEFENCE OF A CONVOY

THE SITUATION The time is the early autumn of 1915 and the scene the Adriatic Sea where, since Italy's entry into the war in May of the same year, fighting has occurred intermittently. The ships involved have been of Italy, France and Britain on the one hand, and of Austria–Hungary and Germany on the other. For the most part the fighting has been on a small scale, but a joint Anglo-Italian force has just inflicted severe damage on the road and rail routes which run along the Dalmatian coast. As a consequence the effectiveness of the Austrian U-boat base at Cattaro (present-day Kotor) has been much reduced, and the Austrian government needs to find an alternative means of supplying the base with torpedoes if the Austrian and German submarine offensive is not to lose its momentum. A convoy is therefore prepared at the northern base of Pola with the intention of sending it overnight down the Adriatic. Because of Allied mining along the Dalmatian coast, it is felt necessary to route the convoy well out to sea, the shelter of the islands west of Dubrovnik only being sought towards dawn when the last lap of the journey would be beginning. A strong screening force consisting of the Austrian pre-Dreadnought battleship *Radetsky*, the German light cruisers *Augsberg* and *Regensberg* and four German destroyers *V25*, *V26*, *V27* and *V28* was to accompany the convoy which at best was expected to make about ten knots. Four Austrian submarines *UB1* to *4* were to leave Cattaro the evening before the convoy sailed to give assistance should Allied surface vessels intervene. Thanks to the shrewd observations of Italian spies in Pola and their promptness in getting a message to Venice, the Italian naval authorities there were able to despatch two vessels only a few hours after the convoy had sailed. These were the battlecruiser *Roma* and the protected cruiser *Libia*. Steaming in excess of twenty knots for most of the night, they had by dawn closed the gap between the convoy and themselves to only a few miles. Meanwhile a British squadron consisting of the light cruisers *Arethusa* and *Dartmouth* and the four destroyers *Mons*, *Matchless*, *Milne* and *Murray* had left Brindisi the evening before to patrol the lower Adriatic. The Austrians had made the habit recently of sending light surface vessels across the Adriatic from Cattaro to shell such Italian cities as Bari, slipping away as quickly as they had come. Aerial reconnaissance reports of the previous night suggested that some Austrian submarines were operating in the area.

The Austrians and Germans hope that three or more of the eight freighters with their precious cargoes will get through. The escorting warships are urged to risk a great deal in order

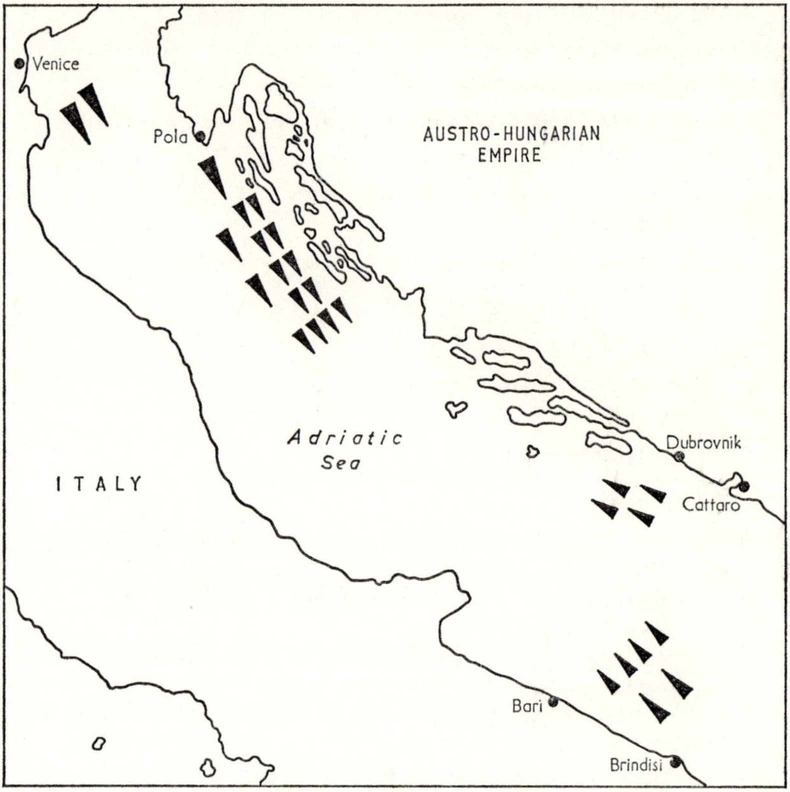

FIG 14 Map of battle area

that this objective should be achieved. For their part, the Allies' ships have orders to destroy or turn back the convoy—the sinking of five or six freighters being considered a fair exchange even if several warships are lost in the process.

CAPABILITIES OF THE SHIPS In the terms of our rules the speeds of the ships are:

Submarines move one square surfaced or submerged.
Destroyers move five squares in the long move, four in the short.

Arethusa, *Augsberg* move four squares each move.
Dartmouth, *Regensberg* move three squares/four squares.
Radetsky, *Roma* and *Libia* move three squares per move.
The freighters move two squares in the long move, one in the short.

WEAPONS CARRIED

12in, 9·4in, 8in guns all fire three squares.
6in, 5·9in, 4·7in, 4·1in, 4in, 3·9in, 3·5in all fire two squares.
The torpedoes fire two squares.
3in guns fire one square.

DETAILS OF THE FORCES

1 Anglo-Italian (sr = sinking rate)
Roma (1908), 12,625 tons (sr 400), two 12in, twelve 8in, 2 torpedoes
Libia (1912), 3,800 tons (sr 110), two 6in, eight 4·7in, 2 torpedoes
Dartmouth (1911), 5,250 tons (sr 130), eight 6in, one 3in, 2 torpedoes
Arethusa (1914), 3,512 tons (sr 100), three 6in, four 4in, two 3in, 8 torpedoes
Mons etc (1915), 1,000 tons (sr 40), three 4in, 4 torpedoes

2 Austro-German
Radetsky (1911), 14,500 tons (sr 370), four 12in, eight 9·4in, twenty 3·9in, 3 torpedoes
Augsberg (1910), 4,350 tons (sr 130), twelve 4·1in, 2 torpedoes
Regensberg (1914), 4,900 tons (sr 130), twelve 4·1in, 4 torpedoes
V25 etc (1914), 812 tons (sr 40), three 3·5in, 6 torpedoes
UB1 etc, 127 tons (sr 10), 2 torpedoes
Freighters—miscellaneous (sr 20)

PROSPECTS FOR THE BATTLE Both commanders could feel, on paper at least, that each had good prospects for victory. The Austro-German force had as its mainstay the very powerful *Radetsky*, whilst the four submarines could be expected to play a decisive role were they to get within reach of the enemy. The

Italian vessel *Roma*, although lightly armed, was of very sound construction and well armoured and would be able to absorb a lot of punishment. The British squadron appeared to have sufficient superiority in speed to enable it to elude the submarines.

MOVE 1 The convoy heads towards the islands with the warships lying close to and on its seaward side. Three submarines which were positioned on the extreme north-west of their patrol area head towards the islands, too, whilst the fourth crosses into the Italian 'half' of the battle area. The British squadron is making its maximum speed to the north whilst the Italian vessels steer a course which will bring them within range of the convoy in the next four or five moves. A dice throw had decided that the Allies should move first, so for move 2 it was the turn of the Austro-German force to start the movement turn.

MOVE 2 *Augsberg* and *Regensberg* are detached by the Austrian commander to investigate the Italian force, the smoke of which is clearly visible. The submarines continue to hold a steady course for the islands, but the British squadron has swung away to the west apparently suspecting that submarine attacks might be imminent.

MOVE 3 The Austrian commander decides to split his surface force. *V26* and *V25* dash on ahead of the convoy, presumably to rendezvous with the submarines, as they could not alone interfere with the British squadron's approach. The flagship *Radetsky* and *V27* and *V28* make a fine sight as they change course and turn towards the converging Italian vessels.

MOVE 4 The Allies show some caution now. *Roma* and *Libia* steer on a course parallel to the convoy, whilst the British squadron still disguises its intentions by heading north, north-west. The initiative is with the Austrian commander as his forces move last. He seizes it by putting *Radetsky* within three squares of *Libia* and thus in range, with his own cruisers kept just out of range. All his destroyers are bunching to protect the convoy from the anticipated cruiser attack. The submarines move slowly and steadily towards the islands where a battle now looks certain to take place.

FIRING *Radetsky* has put herself on the broadside to *Libia* and thus can fire all her 12in guns which are in bow and stern turrets, and four of her eight 9·4in guns. *Libia,* with only 6in guns, is powerless to retaliate. Four dice are thrown for the 12in guns with only sixes to score as the range is the maximum. One hit is made and so seventeen damage points are recorded against *Libia*'s sinking rate of 110. But the 9·4in guns are a disappointment. Four dice are thrown for the first salvo, and all are misses. As the target remains the same for the second salvo, fives as well as sixes now count. This time one of them scores a hit. That means eight more damage points are recorded on *Libia,* a total of twenty-five. As this represents more than 20 per cent of her sinking rate, she loses one 6in gun turret. First blood to the Austrians.

MOVE 5 For the Austrian submarines excitement is mounting. *UB1* is now behind the British squadron and athwart its likely path of retreat, whilst *UB2, 3* and *4* are handily placed as they make their best speeds towards the islands. The Austrian commander shows some indecision now. Keeping *Radetsky, Augsberg* and two destroyers at the rear of the convoy, he detaches *Regensberg* to move up ahead with *V26* and *V25.* He might have done better to harass the damaged *Libia* with both his cruisers and his battleship. The Italians, for their part, are glad of a respite and stay lurking out of range although *Roma* is awkwardly placed behind *Libia* and cannot retaliate effectively if *Libia* is attacked. The British squadron, on the other hand, heads straight for the head of the convoy and enters the dangerous waters of the islands.

MOVE 6 The Allies are obliged to move first, always a nerve-wracking moment in a battle when enemy ships are so near. An over-bold move would see them completely enveloped by enemy ships, an over-cautious one would enable the enemy to avoid combat. On this occasion the Italian ships act cautiously, falling back a little and trying to get the Austrians to spread their forces over an ever larger area. In contrast, the British steer for *Regensberg*, apparently unworried that they are now in waters where Austrian U-boats might be lurking. The British commander later revealed that he had become bored with trying to work out how far submarines might have reached after six game moves—a piece of carelessness which was soon to cost him dearly. The Austro-German ships throw off their previous hesitancy when it comes to their move. *Radetsky, Augsberg, V27 and V28* all go in hard at the *Libia*. The situation is shown in Fig 15.

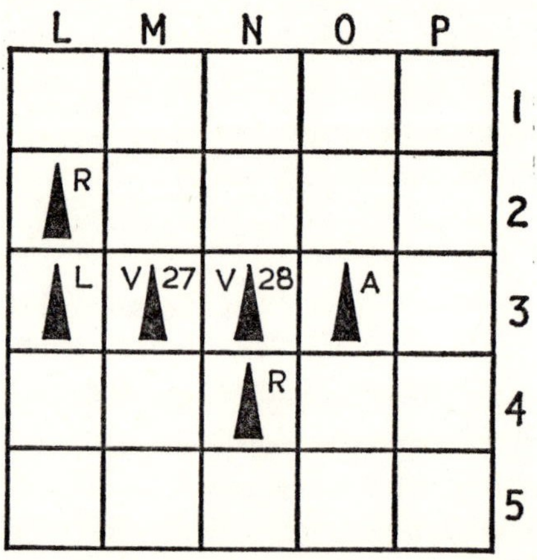

FIG 15 *Roma* (L2) and *Libia* (L3) can fire on both destroyers. *Radetsky*'s guns (N4) can reach *Libia* and so can both destroyers' torpedoes

FIRING *Libia* fires first her one sound 6in gun, firing salvoes at *V27*, recording one hit (two points), and as she can bring four 4·7in guns to bear she has three salvoes with these. Fours, fives and sixes count as hits in the second and third dice throws as she is firing at less than maximum range and at the same target. Four hits are recorded—another two and two-thirds points of damage done. As the ships are broadside on to each other it is well worth trying a torpedo—she only has one on each side of the ship. Two dice are thrown for an aggregate of only six, a miss. So there is not much to show for *Libia*'s efforts. *Roma* chooses the same target, *V27*, and gets a six with one of her two 12in shots. Seventeen more damage points are recorded. She then fires her first salvo of six 8in guns at the same destroyer, and has the lucky result of four hits with throws of 5:5:5:6:1:3. Twenty more damage points on *V27*, a total of forty-one and two-thirds, and sufficient to sink her after her own firing turn. *Roma* is allowed by the rules to switch her second 8in salvo to another target. Her choice is *V28*, but the six dice throws are at maximum range and produce only one six. So although *V28* is damaged, the Italian firing has ended with only one destroyer accounted for. Now *Roma* and *Libia* brace themselves. In fact the Austrians do not even need to use all their fire-power. *V27* fires her guns at *Libia* and then, decisively, fires three torpedoes, of which two are hits and score 140 on *Libia* which sinks.

MOVE 7 The Austrians move first. *Radetsky*, *Augsberg* and *V28*, after their successful aggression, move back close to the convoy. *Roma* shadows the convoy and gets *Augsberg* within range. At the head of the convoy the Austrian and German warships close ranks and wait for the expected British onslaught which now seems imminent as the much higher speed of the latter has put their squadron almost in the convoy's approach path.

FIRING *Roma*'s 12in guns score once on *Augsberg* and the 8in ones four. *Augsberg* loses one turret, and with more than 30 per cent damage is close to losing speed, too. For the Italians this is a useful tonic, so soon after losing their own cruiser.

MOVE 8 The battle seems to be in the balance. *Libia* has sunk, but against that the Austrians have lost a destroyer and *Augsberg* is severely damaged. The Austrians have become split so in this move, and after the British squadron has tucked in behind Cazza Island ready to attack the convoy, the Austrians try to bunch again so that ships will be able to give each other mutual support.

FIRING *UB4*'s patient shadowing of the British vessels is rewarded. *Dartmouth* has slowed down and presents her broadside to the submarine. The submarine's two torpedoes produce one hit—seventy damage points—and the cruiser loses two of her 6in turrets and her speed is reduced.

MOVE 9 The convoy heads more towards the Italian coast in order to minimise the attack of the British squadron. The warships are grouped compactly at the rear and head of the convoy but are unable to prevent the British ships falling on the most exposed freighters.

FIRING *Mons* and *Milne* take a freighter apiece, and gun and torpedo hits quickly account for both targets. Now the whole convoy's fate hangs in the balance.

MOVE 10 There is a flurry of signal flags as the *Roma* joins the destroyers *Mons* and *Milne*, and all seven Allied ships move in on the convoy which is still heading south-west instead of south-east as it would have preferred. Of the six freighters, *F3* is in range of *Matchless* but presents a stern view to the destroyer which turns to use its broadside tubes.

FIRING Only *Matchless* has a stern view of the freighter, and her torpedoes miss. *Radetsky* only hits *Mons* with a 9·4in shell after throwing her full broadsides at the British destroyer.

MOVE 11 At this stage of the battle the accurate firing which has previously been a feature of the Anglo-Italian force seems to desert its vessels. Certainly the closeness to each other of so many fast-moving ships produces a great deal of smoke, but *Roma* fires her 12in and 8in guns at *Augsberg*, damaged earlier, and misses with the lot. *Matchless* again fires torpedoes at *F3*, and misses. More remarkably *Milne* has *V28* as a broadside target but her torpedoes miss. She and *Milne* have now exhausted their torpedoes. Only *Murray* has any success, one of her two torpedoes hitting *Augsberg* which is slowed to two squares' speed and loses further turrets. By contrast the Austrians and Germans fire well. *Radetsky*'s rear 12in turret fires on the damaged *Mons* and completes her destruction. Her 9·4in salvoes damage *Milne*, and *V28* finishes off the British destroyer with a torpedo. *V28* can also fire her other set of torpedoes at *Matchless* and both are hits, sinking the destroyer. Only *Murray* of the destroyers escapes destruction, torpedoes from *Regensberg* and *V25* having missed her.

MOVE 12 The balance has swung dramatically in favour of the Austro-German force. It has only lost one destroyer and three-quarters of the convoy survives. Now the freighters resume their proper course towards Cattaro. *V25* and *V26* are flung at the damaged cruiser *Dartmouth* which manoeuvres to face them thus presenting an oblique angle to *UB3*. Even as the cruiser's own torpedo sinks *V25*, she herself is hit by the gunfire of the destroyers and *UB3*'s successful shot. She sinks. *Arethusa* suffers superficial damage from *Radetsky*'s guns.

MOVE 13 Cattaro is still many game moves away and the Italian commander orders up *Roma* to tackle the convoy's escort. Her guns account for the damaged *Augsberg*.

MOVES 14 AND 15 The Allies trail the convoy. How the situation has changed from the time when the convoy was trapped between two enemy groups! *Murray* alone speeds off on an overtaking course whilst *Roma*'s big guns fail to hit *V28*.

MOVE 16 *Roma* takes a chance and closes in on the freighter *F3*. It is her 8in guns which account for the vessel. But the price is a high one. *Radetsky*'s guns do fifty-eight points of damage, and *UB2* is presented with a perfect view of the target —one of her torpedoes scores a further seventy points. *Roma* limps out of line, a 12in turret out of action, and her own torpedo passes harmlessly astern of *Radetsky*.

MOVE 17 The remaining five freighters bunch tidily, and the protective screen looks as strong as ever. *Radetsky* and *Regensberg* have only slight damage. All the submarines and three German destroyers are intact. The Italian commander, so near to success at one juncture, reluctantly signals his ships away from the convoy and steers a course for Bari, thus dodging the lurking submarines. In a few hours the hundreds of torpedoes carried by the surviving freighters will be unloaded at Cattaro.

In the inevitable post-battle inquest both players agreed it had been a close-run thing, the commander of the Anglo-Italian force feeling that if only his shells and torpedoes had found their mark in move 11 the battle would have had a very different outcome.

GAME 2: THE HUNT FOR THE RAIDER

THE SITUATION When setting up a game with a raiding warship, it is tempting to choose a German pocket battleship as that raider. There is something particularly compelling about the notion of 11in guns being carried on a conventional cruiser hull. But many warships can take on the role of commerce raiders, and in the action described here the lone warship is the Japanese heavy cruiser *Atago*. This ship has stayed in the Indian Ocean after the striking force of the Japanese navy had withdrawn in the spring of 1942 to face the growing might of the United States navy. Supplied by an equally elusive depot ship, it had sunk many merchant ships crossing the Indian Ocean and its presence had led to the setting up of small search groups each, it was hoped by the Allies, capable of tackling the cruiser unaided if necessary. It is now 10 August 1942 and the recent pattern of merchant-vessel sinkings suggest that the Japanese raider is approaching the Nicobar Islands off Malaya, presumably returning to the Japanese-controlled waters to the east. Four of the Allied search groups are near enough to the area to be ordered to converge on the islands. Unless they find the cruiser quickly she will soon be within range of protective air cover based on Malaya.

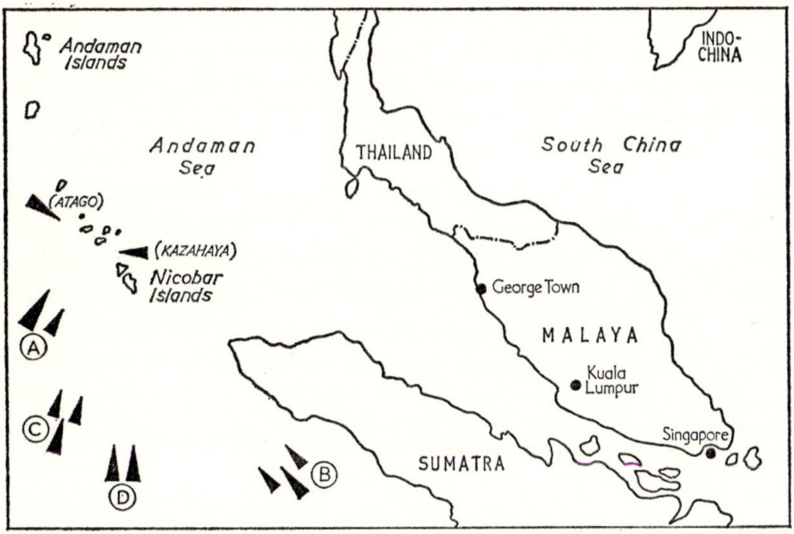

FIG 16 Map of battle area

THE WAR-GAME BOARD AND SPECIAL RULES I concentrate for game 2 on providing the mechanics of the game and only outline the action itself briefly.

The twelve game-board pieces I use were arranged two by two to make a long and narrow playing area twenty squares by forty-eight. Islands were scattered liberally over the entire playing area.

The players were bound by the following factors:

(*a*) The raider remains unseen to the Allies until it is within five squares of an Allied ship or a plane spots it.

(*b*) The raider enters at the northern end of the chart at a point of his own choosing.

(*c*) The raider achieves victory if he leaves the playing area at the far end of the chain of islands.

(*d*) The Allied forces, four in all, are placed on any of the twelve game boards according to dice throws, each board having been assigned a number in advance.

(*e*) Following the placing of the Allied ships, the Japanese player may place his depot ship anywhere he likes in the battle area, again invisible to the Allies.

(*f*) Those ships carrying seaplanes can send them off on flights of twenty-four squares, each consisting of four game moves. Recovery is by stationing the ship on the same square as the plane for one whole game move. A plane 'sees' nine squares as it flies.

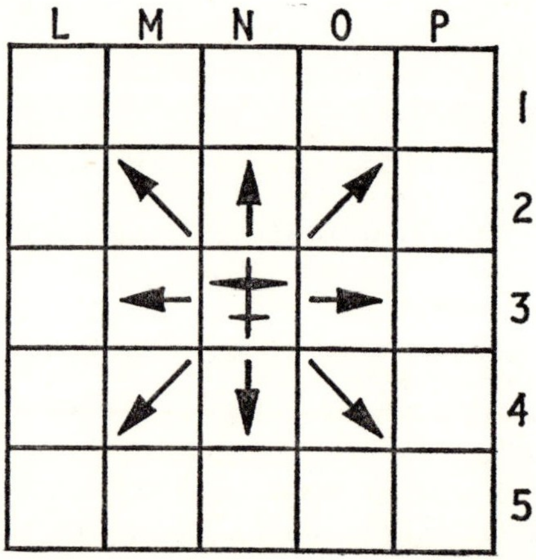

FIG 17 The squares 'seen' by a searching plane

Once either of the Japanese ships is spotted it must be put on the game board and kept on unless it has three consecutive moves in which it is not observed, or has been more than five squares from a ship.

DETAILS OF THE SHIPS

1 Japanese (sr = sinking rate)
Atago (1930), 13,160 tons (sr 200), ten 8in, eight 5in, 16 torpedoes
Kazahaya (1942), 18,300 tons (sr 50), four 5in, three 4·7in
Atago was capable of 36 knots, *Kazahaya* of 16.

2 The Allies
Group A British
Royal Sovereign (1916), 29,150 tons (sr 600), eight 15in, twelve 6in, eight 4in, 1 plane, 21½ knots
Electra (1934), 1,375 tons (sr 55), four 4·7in, 8 torpedoes, 35½ knots
Group B Free French/Dutch
De Zeven Provincien (1947), 10,000 tons (sr 180), ten 5·9in, 6 torpedoes, 2 planes, 32 knots (a post-war vessel)
Le Hardi (1939), 1,982 tons (sr 60), six 5·1in, 7 torpedoes, 35 knots
Le Corsaire (1941), 1,982 tons (sr 60), guns, torpedoes, speed as *Le Hardi*
Group C American
Biloxi (1942), 10,000 tons (sr 160), twelve 6in, twelve 5in, 4 planes, 33 knots
Radford (1942), 2,050 tons (sr 55), five 5in, 10 torpedoes, 37 knots
Jenkins (1942), 2,050 tons (sr 55), guns, torpedoes, speed as *Radford*
Group D New Zealand
Achilles (1932), 7,000 tons (sr 160), eight 6in, eight 4in, 1 plane, 8 torpedoes, 32 knots
Leander (1931), 7,000 tons (sr 160), guns, planes, torpedoes, speed as *Achilles*

Note: Group A was built around the elderly battleship *Royal Sovereign* which, because it had not been modernised, could only fire its 15in guns at the same range as 8in gun cruisers. The other groups consisted of cruisers and destroyers.

STARTING POSITIONS OF THE SHIPS The dice throws produced the following disposition of the ships as shown in Fig 18.

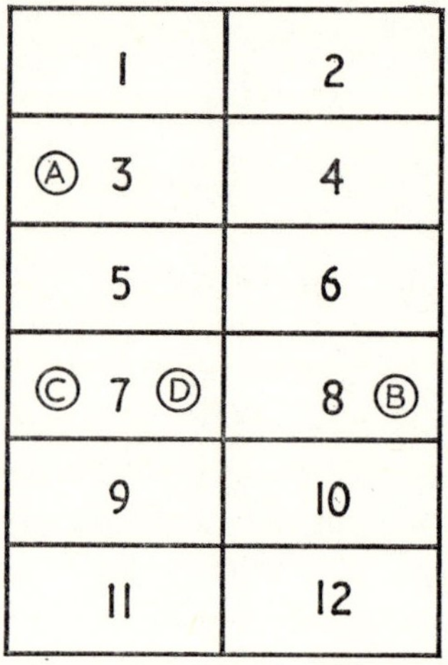

FIG 18 The raider chose to enter on 2. The Allied squadrons A, B, C, D found themselves as above. The Japanese depot ship was placed on 10

After the Allied groups had been placed on the game board, the Japanese commander chose to place his depot ship on area 10 in the hope that it would be able to render assistance if *Atago* got that far through the islands.

SUMMARY OF THE BATTLE The battle's outcome was decided after fifteen moves. An aircraft of the *Biloxi* spotted *Atago* in game move 3 as the Japanese cruiser headed boldly down the centre of the chain of islands. Thereafter she was never to be out of sight again although the depot ship escaped detection. It soon became apparent that *Atago* could never really hope to throw off her pursuers but would have to sustain some damage by charging past one hunting group and hope to keep faster vessels at bay with gunfire as she dashed southwards.

Consequently *Atago* headed for the New Zealand cruisers and only shied away from action with them when *Biloxi*'s group moved westwards in support. Thus the Japanese ship found herself heading due west with groups C and D blocking her route to the south and group A to the north with the *Royal Sovereign* just out of range. *Achilles* and *Atago* exchanged shots and scored usefully on each other. *Atago* now took a gamble and headed back to the north-east, sinking the destroyer *Radford* and dodging *Electra* which had stopped to pick up the pilot of *Royal Sovereign*'s ditched aircraft—and the gamble paid off for *Royal Sovereign*'s complete salvo of 15in shells missed the cruiser.

Atago's respite was short-lived. Her new course took her within range of the cruisers *De Zeven Provincien* and *Biloxi* and whilst both scored gun hits, the latter did particularly well and *Atago* lost an 8in turret. Although she had by now suffered sixty-seven points worth of damage, her own 8in shells left *Electra* crippled and ablaze.

Both players felt that the result was now never in doubt but *Atago* gave the Allies a good chase for their money largely because, it seemed in retrospect, the Allied commander was trying to minimise casualities by trying to sink *Atago* with long-range gunfire. This policy left a 'gap' through which *Atago* dashed in move 8 and although she was once again mauled by *Biloxi*'s fierce and accurate gunfire, she sank the American cruiser with torpedoes. By now she had 62 per cent damage, had lost three 8in turrets and, most crucial of all, had lost a square of speed. The remaining moves saw *Atago* heading due south pursued by all the remaining Allied ships except for *Electra* and the damaged *Radford*, and in the unequal gunnery duel *Atago*'s damage mounted. Her own guns inflicted damage on the *Achilles* and *De Zeven Provincien* and she caught *Jenkins* with a salvo that sank her, but her firing became more and more fitful and her speed dropped to less than fifteen knots. The *Leander* joined the Dutch cruiser to finish off the barely moving enemy and despite the fact that *Leander* put herself within range of the Japanese Long-Lance torpedoes, these could not be fired because of the cruiser's heavy list (96 per cent damage).

INQUEST ON THE GAME The game proved to be one of those

which leaves both players arguing and discussing for a long time afterwards. The main issue of contention was whether or not *Atago* really ever had a chance to escape. The Allied commander certainly felt he could have crippled her with destroyer attacks much earlier than he did had he not been trying to avoid losses which would have been inevitable in such attacks. The Japanese commander argued that the playing area was too narrow to allow him to use his speed to greatest effect, and as he could not outrun his pursuers his 8in guns gave him no particular advantage. He added that if only he had had 11in guns the story would have been different!

Both players enjoyed the early hide-and-seek moves, the chase and the unavailing gunnery action of the Japanese cruiser; and the Japanese player could point to the fact that his ship had sunk a cruiser and a destroyer, severely damaged two other destroyers, and damaged two more cruisers. Honour had been achieved.

CAMPAIGN SUGGESTIONS

A campaign, for naval war-game purposes, is a series of battles which collectively decide the outcome of the war. Through having to fight more than one battle, a player will clearly need to allocate his forces for each encounter with an eye to his likely casualties and their bearing on his strength for the next engagement. On some occasions, therefore, he may prefer to make only token resistance in order to husband his resources for what he considers are the really crucial occasions. Or the campaign might be so organised as to give each player a wide choice of objectives in pursuit of which battles inevitably occur. As far as the game side of a campaign is concerned, the players involved will need to do much more preparation than for a solo battle but the research and rule-making involved is just another facet of the interest the hobby provides.

I now describe the elements which went into the planning of one campaign with which I have been concerned. It will readily be seen that features such as the consumption of fuel and the part played by land warfare can and should be introduced for a campaign. But I again assume that the war gamers tackling this campaign will not have an umpire so it is not possible to have a great deal of hidden movement. I do not give the result of this particular campaign. Instead I hope the reader will want to fight it himself, or at least one like it. I chose to base this campaign on an imaginary clash in the Far East towards the end of World War I. This theatre of war provides a compact area for naval warfare with scope for land operations and naval actions in support of them. The Mediterranean Sea, the Baltic Sea and the Caribbean Sea also suggest themselves for campaign theatres. Or one may create an entirely imaginary part of the globe and people it with mythical nations.

THE WAR FOR KOREA 1918 China and Japan have once again clashed over who should exercise influence in the republic of Korea (we alter the facts here too!). On this occasion the Korean government and people have come out in support of China and had matters stood there Japan would probably have done nothing more than send stiff diplomatic notes to Peking. But Russia, out of the war in Western Europe but with a large army in the Far East, has taken the opportunity to side with Japan. Her justification for declaring war on China is that the Chinese keep interfering with the passage of goods and personnel through Manchuria to the Russian-held port of Port Arthur. The three major nations are now preparing to move armies in support of what they regard as their justifiable claims. Korea seems to be the likely battlefield for all three.

THE BATTLE AREA A small-scale map of this part of East Asia was copied and the grid pattern (Fig 19) was drawn over it. Each square, according to the scale chosen, should represent sixty miles by sixty. Vladivostok becomes the furthest point north and east, Shanghai the furthest south and Peking the furthest west. The east-west side of the map measures 720 miles, the north-south side 600. One of the longest sea voyages possible is Vladivostok to Shanghai, about 1,000 miles. On the other hand, the distance between the Japanese port of Shimonoseki and the Korean one of Pusan is only 135 miles.

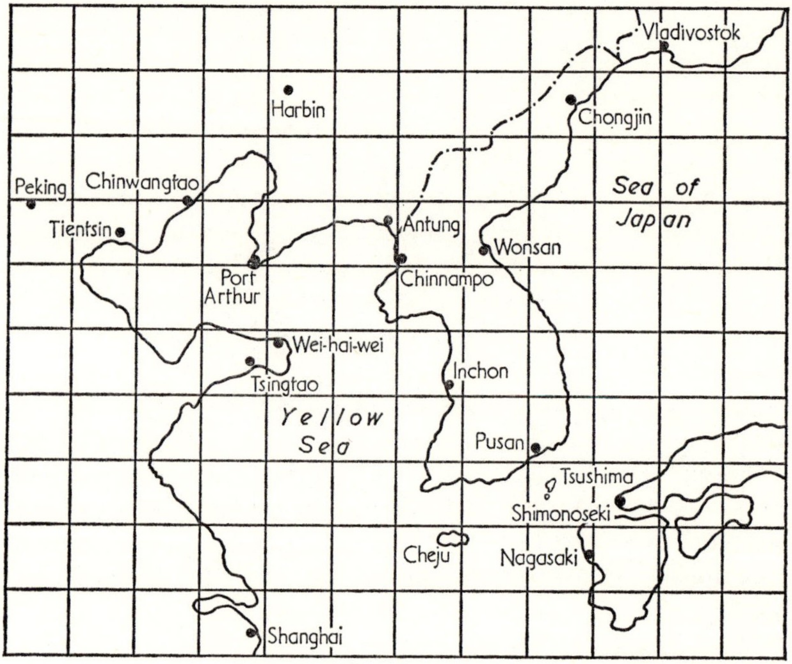

FIG 19 The East Asian area chosen for the author's campaign game

PORTS Only the ports listed below may be used during the campaign. All can provide repairs, fuel, embarkation facilities and munitions except the island ports of Cheju (Korean) and Tsushima (Japanese), which do not have repair facilities.

China	*Korea*	*Russia*	*Japan*
Antung	Cheju	Port Arthur	Nagasaki
Chinwangtao	Chinnampo	Vladivostok	Shimonoseki
Shanghai	Chongjin		Tsushima
Tientsin	Inchon		
Tsingtao	Pusan		
Wei-hai-wei	Wonsan		

Each port may have up to 2,000 units of fuel for shipping provided that the totals for each nation do not exceed the following:

Nation	*Fuel units*
China	4,000
Korea	1,000
Russia	2,000
Japan	6,000

Each port has a shore battery containing six 9·2in calibre guns. Only guns of 8in calibre or above can damage these at the rate of twenty damage points for each gun knocked out.

Those ports with repair facilities can repair damage to a ship at the rate of 10 per cent of the ship's sinking rate per game move.

Troops may be embarked or landed at a port at the rate of two moves for each five troop units.

For an army to take a port or a city requires a superiority in troops of three to one. Only then may enemy ships enter the port and use its facilities. If a city is captured, all its remaining fuel units and shore-battery guns are handed over intact to the victor.

A port can accommodate up to 100 troop units.

CAMPAIGN MOVES AT SEA Each game move at the campaign level denotes six hours of real time. Thus ships may be moved from one campaign chart square to another each move if their speed is ten knots (six hours equals approximately sixty miles travelled). At higher speeds greater distances may be covered but with a corresponding increase in the amount of fuel used up.

Fuel consumption is assessed on the basis of a ship's likely cruising speed—that is, the highest speed giving the most economical consumption of fuel.

The cruising speed for any ship is regarded as three-fifths of its maximum speed. At this speed the ship uses up x units each game move of six hours. The table below shows how fuel consumption then varies with speeds lower or greater than the cruising speed.

Speeds (at fractions of maximum)	*Fuel units consumed*
$\frac{1}{5}$	$\frac{1}{2}x$
$\frac{2}{5}$	$\frac{3}{4}x$
$\frac{3}{5}$ (cruising speed)	x
$\frac{4}{5}$	$2x$
Maximum	$4x$

X varies according to the type of ship, so does the storage capacity for each ship. Both are given for a sample range of ships.

	x *(units)*	*Storage capacity (units)*
Submarine (on surface)	$\frac{1}{4}$	8
Troopship	$\frac{1}{2}$	16
Destroyer	1	9
Cruiser	2	26
Battleship pre-Dreadnought	3	39
Battleship Dreadnought	4	48

Troopships are deemed capable of speeds of sixteen knots, but cruisers are restricted to ten knots.

Submarines moving underwater are not affected by fuel consumption since they can recharge their batteries periodically.

Ships may refuel at sea with units up to their storage capacity, taking one game move alongside the depot ship to do so. Fuelling in port takes the same length of time.

A ship running out of fuel at sea is deemed stationary and cannot alter its position when attacked.

COMBAT AT SEA Players move their naval forces on the main campaign chart in view of their opponent but with the exact composition of a force only being revealed when forces of opposing sides are in the same campaign square. Then combat may occur and all the ships of the opposing forces are placed on the main battle chart. I used all twelve battle boards (see Chapter 4) giving an area of 32 by 30 squares. Any features such as islands and harbours should then be added by mutual consent, and the models depicting each ship should be placed on the battle board on the side appropriate to their entry on to the corresponding campaign square.

For combat purposes each game move now represents half an hour, so only after twelve combat moves has six hours elapsed and at that point attention must be given to all the other land and sea forces on the campaign chart. Some of these, of course, may be moving up to take part in the combat themselves. Ships may move at maximum speeds during combat, but after each twelve combat moves the fuel consumed for a cruising-speed move is recorded.

Combat is exactly as in the rules outlined in Chapter 4 and ends when either player takes his ships off the edge of the battle area.

USE OF TROOPS Five units of troops may embark in any one troopship. They are lost entirely if the ship is sunk. On land they are regarded as self supporting for forty campaign moves. If a unit is not then replaced, its fighting capacity is halved.

Troops may embark or land using port facilities at the rate of two campaign moves per five units. To land or disembark away from a port takes four campaign moves per five units. If troops are opposed when landing they must have three to one superiority to be successful. Fifty troop units may occupy one campaign-chart square. On land, troop units move at the rate of half a campaign square (thirty miles) per game move (six hours) and then rest for one game move. If they choose to move sixty miles in two successive game moves then their subsequent rest is two game moves (twelve hours). Players do not reveal the composition of any group of units until combat occurs.

Combat occurs when troops meet on the same half-square. When the odds are two to one in favour of the force actually moving, then victory is achieved and all the opposing units are removed from the chart. If the defenders are in a city, then odds of three to one are necessary for the attackers to succeed.

The four nations start the campaign having the following number of troop units: China 200; Korea 20; Russia 100; Japan 80.

We have already seen that whilst any campaign square can hold up to fifty units, a port can hold up to 100 units. It can be readily seen that Russia with two ports and 100 units can place all her troops to begin with at Port Arthur or Vladivostok.

It is suggested that a nation should only have a limited number of troopships. This gives an incentive to each nation to sink them on the one hand, and protect their own on the other.

Nations therefore have the following number of troopships: China 60; Korea 6; Russia 30; Japan 32.

Japan has the greatest proportion of troopships to troop units because none of her troops could fight in Korea without being transported by sea.

ALLOCATION OF SHIPS This can be left to the players concerned and will depend on the amount of time they wish to devote to the campaign, the number of models they have and so on. But I recommend that whatever yardstick is adopted—for example, the Ship Value Formula introduced in Chapter 4—the proportions of vessels per nation should be as follows:

China	$2x$	Russia	x
Korea	$\frac{1}{2}x$	Japan	$3x$
	$2\frac{1}{2}x$		$4x$

The Russo-Japanese forces have a superiority of eight to five in order to enable them to counter the Sino-Korean superiority in troop units and their advantages in having a shared land frontier.

KEEPING DETAILS The usual record sheet as used in single battle games is sufficient for this purpose, provided fuel consumption is also noted.

VICTORY CONDITIONS As in real life these need not be known exactly in advance. We have said originally that Japan and Russia wish to control Korea whilst China will be fighting to prevent this. But the Russo-Japanese forces may adopt a strategy which involves knocking out China's fleet and occupying all her ports and not attempting to invade Korea at all. If they were successful China would presumably sue for peace and agree not to aid Korea any further. Again, the campaign may go in China and Korea's favour with Vladivostok, Tsushima and Port Arthur all captured, at which point the Russo-Japanese alliance might agree to halt their aggression and not threaten Korea further. Players will soon see the scope for different kinds of strategy in this campaign. If they have played the board game 'Diplomacy' they may even attempt to get a nation to change sides . . . but we are getting a long way from our naval war games.

SOLO GAMING

Some readers of this book will be keen to start gaming but have not yet found congenial opponents or are not within easy reach of an existing club. Other war gamers will have had my own experience of moving home and job and thus losing regular war-game opponents in one fell swoop. Is it possible to keep up one's interest in the hobby in such circumstances?

One consolation in these situations is that the lone gamer is under no pressure as regards his reading, modelling and planning. He can devise his solo games at leisure and fit the play in entirely at his own convenience. Moreover, if he absolutely must have an opponent to complete his pleasure there are a number of postal campaigns conducted by board-game enthusiasts, news of which appears in hobby magazines.

The element which the solo player misses most when devising a game is the presence of another person making decisions quite independently of his own, and it must be admitted that none of the devices suggested for solo gaming ever quite overcome this lack. The player may be utterly scrupulous in the decisions he makes for each side in the battle and always take what logic suggests is the most advantageous course of action. But it is this very honesty which may rule out the surprise action on one side's part which makes any kind of game situation exciting.

However, the lone war gamer can give special prominence in his pre-game arrangements to hidden movement and to surprise changes in the battle situation and thus overcome this lack to some extent. I would suggest in the first instance that groups of ships be assembled with a points value as near equal as can be achieved. Each side might have five groups, each numbering 10,000 points in value. A typical group would then consist of an aircraft carrier and a destroyer, or three pocket battleships or one cruiser and six or seven destroyers. These groups should be given a letter, A to E, and these are written on five identical pieces of card for each side. They are shuffled and one is removed by blind selection so that each side is left with four pieces of card blank side up. Already an unknown factor has been introduced. The group omitted from the pack *may* have been the only one with an aircraft carrier in it, or it may have included most of the destroyers chosen for that side. Then eight more wholly blank pieces of card, identical in size and colour to the lettered ones, are shuffled in with each side's little pack. Given that the game situation involves each side starting its movement in one half of the game board, then the gamer divides up each half of the board into six zones allocated the numbers one to six. He then throws a die six times and depending on the score each time, he places two of the cards in the middle of the appropriate zone. In the unlikely event of him throwing the number six times all twelve cards would, of course, end up in the same zone. More likely is that two cards will be assigned to several zones with others remaining blank and that one zone will end up with four cards on it. The player then apportions the other side's cards in the same manner.

For the first two game moves the player moves all the groups of cards, but at the slow speed of two game squares per move. At the beginning of the third move he turns over the cards of one group of units for each side and will thus know if the group concerned consists of any ships at all. The knowledge gained will probably affect subsequent game moves since the turning up of wholly blank cards will alter a player's capabilities in a particular zone of the game board. At the beginning of the fourth move, the sixth move, the eighth move etc, a group of units should be revealed in the same fashion. At the same time any units must reveal their composition when the other side has a group of units definitely including ships within eight squares of them. Similarly, the presence of a spotter aircraft, or a full aircraft strike against an as yet undeclared group of cards, obliges the player to turn the cards of the group over and reveal the exact composition of the force. As the game proceeds the unknown element is entirely removed, but in the meantime battles will have occurred sometimes in circumstances which have favoured one side more than the other—a situation common enough in real warfare.

This element of concealment can be reintroduced by having a period of night after, say, every eight game moves. Any models on the game board are removed and replaced by a card corresponding to each ship. These are turned face down and in the subsequent two moves representing night and in each zone having ships in it, the cards are divided into two groups and these must move at least four squares away from each other by the time the first morning move comes round. Such movement may have been dictated by the need to confuse searching submarines and will certainly provide the other side with problems when daylight comes. In the first two morning moves, two groups of cards are turned over and replaced, if necessary, by models. As for the idea of having blank cards at all, it is easy enough to rationalise the use of such a ploy by remembering that inaccurate intelligence reports and mistaken identification reports by both ships and aircraft often leave an enemy's whereabouts in doubt.

An attractive way of relating the solo battle to larger strategic considerations is to have a series of strategy cards from which each player draws before, say, the second, fourth, sixth etc move. These may order the commander of a convoy escort to release one warship and send it off after a reported submarine; again, a commerce raider may find itself ordered to alter course and to proceed to another game-board zone in the expectation of finding victims there. Yet again, a ship might be ordered back to base for a refit. These will appear arbitrary decisions, but such turns of events are not unknown in warfare.

Another device, and one which is equally appropriate where more than one player is involved, is to assign points values to shore installations such as factories, bridges, reservoirs, harbour installations, shore batteries, railway lines and canals, but only to signify the nature and value of the target on small cards which are shuffled and placed face downwards on the game board's shore area. A player may only know what the target is by sending an aircraft over it or by landing a shore party from a warship like a submarine or motor torpedo boat. He will be seeking to destroy so many points' worth of shore installations whilst his opponent's task is to stop him.

I hope I have said enough to indicate that the solo player can give himself a reasonably exciting game.

Chapter 6

READING ABOUT NAVAL WAR GAMING

It is natural for the newcomer to naval war gaming to seek out all the books and journals devoted to the hobby so as to get the benefit of other people's experience in the field. There are two books which any naval war gamer would do well to study. The first, *Naval War Games—Fighting Sea Battles with Model Ships*, written by that indefatigable war gamer, Donald Featherstone, appeared in 1965 and remains in print. In it Mr Featherstone gives much useful advice for battles from ancient times to World War II. The book has six sections: games using model ships, model ships for war games, rules for fighting naval war games, pioneers of naval war gaming, operational choices, and suggestions for further reading. As well as his own practical knowledge, Mr Featherstone draws upon the experience of many war gamers who have corresponded with him over the years through his journal *Wargamers' Newsletter* and so there are many well-known enthusiasts here: Tony Bath, Jack Scruby, Charles Dick, Joe Morschauser amongst others.

For the 1914–18 and 1939–45 periods there are various suggestions for rules with some ingenious proposals for concealing one fleet from the other for as long as possible; the major part of Fletcher Pratt's *Naval War Game* rules are reproduced and there are descriptions of battles fought out by Mr Featherstone and his friends. I had come to naval war gaming before this book was published but from the first found it useful and stimulating. The reader may even find it converts him from the modern era to the colourful days of sail—and from my point of view that would be a pity!

P. Dunn, author of *Sea Battle Games* (1970), concentrates on providing rules and hints for fighting battles in four periods—the Napoleonic Wars, the 1870 ironclads period, World War I and World War II—although there is also advice for those wishing to reproduce battles of about 1900. The detailed explanation of rules is supplemented by hints on buying or making model ships, on the role of maps in war gaming, and suggestions for fighting a campaign. The book is well illustrated with photographs and helpful diagrams, and whatever set of rules one eventually formulates for a game, there will always be value in studying Mr Dunn's examples which are based on careful research into the ranges of weapons, endurance of warships and the performance of the ships in the real thing.

Two other books have some value, although their purpose is more general. They both come in the admirably cheap and attractive Shire series. The first, Arthur Taylor's *Rules for War Gaming* (1971), includes a few pages on modern naval warfare with a suggested set of rules. If one adopts these rules, Shire Publications sell pads of war-game stationery which save preparation time when setting up and fighting a battle. The other, *Discovering Ship Models* (1971), written by Norman Boyd, is above all useful for its selective list of museums in Europe and the USA where ship models can be seen.

For the war gamer who wonders how simulation exercises are carried out at the highest level, for example in the armed forces and government agencies, Andrew Wilson's book *War Gaming* (1970) will provide food for thought. Here are descriptions of many war games played by military high commands from the mid-nineteenth century to the late 1960s. One learns, for example, that the United States Navy was staging games at a high level of realism from 1922 onwards and had tried to anticipate every possible facet of a Pacific War. Admiral Nimitz was to say in 1950 that such war games had reproduced at one time or another everything which the Japanese were to attempt between 1941 and 1945, except the use of kamikaze tactics. Andrew Wilson also shows that professional war gamers are not averse to the sort of cheating which sometimes occurs on the table top. He quotes from the book *Midway*, by two former Japanese naval officers, Mitsuo Fuchida and Masatake Okumiya, the description of the war game held aboard the battleship *Yamato* in May of 1942 as a prelude to the Japanese attack on the Midway islands. The throwing of dice determined the success of air strikes against ships and this resulted in the Japanese fleet losing two of its vital four aircraft carriers. However, the presiding umpire, Admiral Akagi, promptly waived one sinking and at a subsequent stage of the game the other lost carrier reappeared. Other aspects of the game proceeded smoothly for the Japanese forces as Admiral Akagi overruled several unfavourable decisions of the umpires. In the subsequent action of 4 June, all four Japanese carriers were lost as against the one lost by the Americans and they were destroyed by a flank attack which the war gamers had neglected to build into their simulation.

Although Andrew Wilson's book is a source of ideas it is also a warning against making easy assumptions that even the most sophisticated simulation exercises can cater for all the variables of the real thing.

All the books I have mentioned are recent ones. How did earlier generations of war gamers cope with simulating naval warfare? The father of the hobby must surely be John Clerk who published his *Essay on Naval Tactics* in 1790. The book can only be consulted in copyright libraries such as the British Museum and is of course of direct value only to the sailing-ship war gamer. But it is interesting to read that Clerk and a friend studied the despatches of actual battles, like the action involving Admiral Byng off Minorca in 1756, and then tried to refight them with model ships. He claims that the use of models removed the possibility of disagreement when he and his opponent were trying out all possible combinations of actions in a battle. It is also touching to know that Clerk carried those models with him constantly, 'every table furnishing sea room sufficient on which to extend and manoeuvre the opponent fleets at pleasure'. John Clerk, then, gives our game a healthy pedigree, but it is to Fred Jane that we must turn for rules which can have application in the modern era.

Fred Jane is now best remembered for starting the magnificent series of reference books on the world's navies and which still bear his name: *Jane's Fighting Ships*. These are of course invaluable works of reference to a war gamer and, happily, reprints for some years are now available, including the first one of 1898 and those for 1914, 1919, 1931, 1939 and 1944–5 which will be much consulted by the modern war gamer. But Jane also published in 1901 his *Hints on Playing the Jane Naval War Game* and one learns from the book's preface of frequent games at his home in Portsmouth to which naval officers particularly were invited and in which 'the close simulation to real warfare is the sole objective of the game'.

Apart from detailed rules for all aspects of warfare at about the turn of the century, the book makes valuable suggestions for conducting a campaign. Mr Jane and his friends carried out a major campaign during March, April and May of 1900 and the seventy-five players involved were able to read intelligence reports which the Portsmouth newspaper, the *Southern Daily Mail*, obligingly published each day. In this campaign weather, the unexpected breakdown of ships, false intelligence and disguised ships all played their part. Coal consumption was allowed for, a ship at top speed using fuel at four times the rate of its lowest speed. Ports accrued points if merchant ships avoided the blockade, but if the blockade was successfully maintained then ports lost points and thereby the means to provision ships etc.

The definitive version of *Jane's Rules* appeared in 1912 and by then his publishers, Sampson Low, Marston & Co, were also selling a full range of model ships—£40 would buy practically all the world's warships—as well as war-game stationery. This set of rules ran to ninety-one pages and had been updated to the extent that submarine warfare was introduced as was the use of balloons and even aircraft. Unfortunately for the war gamer, *Jane's Rules* are out of print.

When we turn to World War II, one man's name heads all others, the American, Fletcher Pratt. A writer on many subjects, but especially naval history, he began devising naval war games in the 1930s when he and his New York friends tired of poker games. Their early rules produced too many anomalies such as single destroyers sinking battleships by gunfire but after many trials over the years a more or less definitive set of rules was evolved. These were published under the title *Fletcher Pratt's Naval War Game* (1940) by which time the real Battle of the River Plate had been fought and the result had confirmed what one of Pratt's earlier war games had suggested, that the pocket battleship *Graf Spee* could be worsted by cruisers. When assessing these rules it has to be borne in mind that Pratt's battles attracted many participants, and very large rooms and even halls were needed in which to play them—panels of umpires kept the action proceeding smoothly. But adherence to the basic principles of his rules would help any newcomer to the hobby.

Turning now from books to magazines, we find an encouraging picture. Two magazines in Britain cater solely for naval war gamers and the newer one is exclusively for twentieth-century war gamers operating in the 1/1,200 and 1/1,250 scale. The newer journal, *Broadside*, appears in January, April, July and October each year and is under the editorship of Mr A. C. James. Among the services it offers to readers are news of commercial models and detailed reviews of some of these, an answering service for enquiries of all sorts, reduced prices for models and books, advertisements of second-hand models and articles about war gamers' activities. The editor also supplies the whole range of Delphin and Miniships. He has in addition been negotiating for facilities for regular large-scale battles for its readers in England, which will probably be held in the Home Counties.

The second magazine, *Battlefleet*, has been produced for

several years and appears approximately monthly. Modestly priced, it fulfils two main functions: to convey detailed instructions to members of the Naval Wargame Society (UK) who battle regularly in a Community Hall in Hook, Surrey, and at a London public house, and to inform readers from further afield of developments within naval war gaming. Full details of the Society's sets of rules and Navwar models are included regularly as is news of all other war-gaming models on the market.

Other general British war-game magazines do occasionally carry items of interest to naval enthusiasts, notably *Wargamers' Newsletter*, *Military Modelling* and *Miniature Warfare and Model Soldiers*.

Of non-British naval war-game magazines, I have not come across any that are widely distributed other than the quarterly *Alnavco Log* which is published by Alnavco Incorporated of New Jersey, USA, the makers of the Superior range of model ships. The *Log*, as one might expect, gives full coverage to new issues of Superior's ships but also publishes many detailed articles about World War II actions or actions fought by war gamers. One enjoyable correspondence which spanned several issues was a debate as to whether or not USS *Iowa*, the largest Allied warship during the war, could have defeated Japan's superbattleship, the *Yamato*, in a straight ship-to-ship duel. The answer can only be a guess since it was carrier aircraft which sank *Yamato* and the sister-ship *Musashi*, but the correspondence did enable one to glean a great deal about the ships' merits and capabilities. Other good features of this little magazine are book reviews and a directory of war-game clubs in the United States.

An American magazine which may eventually command a nationwide readership is *Swabbers*, which is the journal of Michigan Naval War Gamers but which has articles and services of a more general interest to naval gamers.

Finally, but not specifically for naval gamers, *Signal* and *Warship International* are very helpful, the former for news of new products, the latter for authoritative articles on all aspects of naval history.

Fuller details of all books and magazines appear in the notes for this chapter.

Chapter 7

COMMERCIAL GAMES

This book has been concerned so far with helping the war gamer who wishes to fight battles with model ships. But some naval enthusiasts prefer to wage war with pencil and paper and without models. Such players have generally purchased one or more of the commercial board games which simulate naval warfare, the best known of which I describe later in this chapter.

The board game offers several advantages to its users. It requires little space in which to be played; its playing pieces are easily stored and are instantly ready for use; really large-scale battles can be fought with reasonable realism; and the cost of the dearest game compares favourably with the cost of building up a fleet of commercially made model ships. It is even possible to buy blank playing pieces with which to design one's own game. What is missing, of course, is the satisfaction of acquiring and using replicas of warships. Perhaps for this reason the converts to naval board gaming are as likely to come from the ranks of land board gamers as from naval players using model ships. But in fact there is no particular reason why one should not enjoy both types of naval game.

I have grouped the games described in three broad categories according to the extent to which they succeed in simulating naval warfare and enable the player to exercise some skill. These three categories I call simple, standard and complex. The notes (pp 129–31) for this chapter gives full details of the manufacturers, together with a selection of suppliers.

SIMPLE GAMES

All three games in this category are really only variants of the same game, and the basic idea used is such that it is perhaps flattering to call them war games at all. I refer to the commercial productions of the long-popular schoolboy game of 'battleships'. In this each player marks in on a hidden, squared chart the position of his warships and he and his opponent take turns to call out numbered squares until they locate part of a ship when a hit is recorded. The winner is the player who sinks all his opponent's ships. No skill is required although a certain amount of cunning can be exercised in placing one's ships and one may baffle a new opponent for a while.

The three games have some differences, however. The game 'Battleships' by Waddingtons most closely resembles the do-it-yourself version of countless schoolboys. It is really only a pack of ready-ruled stationery, but the usual rules are augmented by the provision of a 'block-buster' shot which, used once in a game, may enable a player to sink a ship covering several squares at one fell swoop.

The version by Merit, also called 'Battleships', has more elaborate equipment, each player having a plastic peg-board into which he inserts markers so that he knows exactly where he has asked when searching for his opponent's ships.

'Bataille Navale' by CEJI is very much more expensive since it comes in the form of an electronic kit which, when assembled, allows the players to call out their opponent's squares by pressing a button.

It can be seen that these games bear little resemblance to naval warfare, yet it has proved difficult in even the most sophisticated games to find a more satisfactory way of searching for enemy warships than calling out squares by their code numbers.

STANDARD GAMES

More ambitious are the games 'Admirals' (Pallitoys) and 'Les Grands Amiraux' (Miro) which, despite their different countries of origin and different packaging, appear to be essentially the same. Two or four players may participate although the additional two players serve only as assistants for the others. Each side places cards representing types of warship on its half of the simple squared board. Some of the cards represent mines and cannot be moved, but the skilful player will also normally leave some warships stationary for as long as possible in order to sow doubts in his opponent's mind as to the exact whereabouts of the mines. The mines can destroy any vessel except a minesweeper, so they pose a deadly threat to the enemy and inhibit his movements until their position is known. The type of ship represented by each card is not revealed until an opponent's card is on an adjacent square and one of the players has said 'attack'. The ship with the higher value wins that battle and the other is removed. Each player has an Admiral card which, if used offensively, defeats all other types of ship. If used defensively, however, it can be taken by even the humblest vessel and its capture does, in fact, signal the defeat of that player. So whilst it is tempting to go on the attack with the Admiral, if it becomes trapped in the process and forced on to the defensive then total defeat follows swiftly.

Ship movement is of the simple one-square-per-turn type, but one ship may sometimes find itself able to hop over several enemy pieces, whatever their value, and defeat them in the manner of draughts.

These simple rules make for a varied and enjoyable game with a certain amount of skill being exercised in the movement of the vessels and in protecting the Admiral card. But it is difficult to reconcile the multi-attack move with real practice.

'Dover Patrol' is a well-tried favourite amongst these elementary but enjoyable naval games, as is testified by the fact that it has sold steadily since the 1930s. The name 'Dover Patrol' is, to my mind, misleading since the game in no way resembles the activities of the group of warships operating in the English Channel and North Sea in World War I.

The object of the game is to capture the enemy's flag and convey it to one's own base square. Capturing involves sinking the ship carrying it or taking it from its base. The board is a simple affair with ninety-six squares of which six on each side are harbours. Since each player commences the game with thirty-six ships, three mines and an aircraft and each occupy a square, the fighting soon begins. Movement is for one piece to move one square each turn. Combat is resolved by comparing the values of opposing ships. The flying boat has advantages which make it very suitable for capturing the flag, but is quite vulnerable to the numerous small warships once its position is known. Thus in several respects 'Dover Patrol' closely resembles 'Admirals' and 'Les Grands Amiraux' although it would be fairer to say that, as the newcomers, they resemble it! The rule book is a model of clarity and includes suggestions for a shorter game and hints for the most advantageous placing of ships at the start of play. Skill can be used in grouping ships to the best effect, in memorising the pieces the opponent has had to show, in placing mines for maximum protection, and in striking out for the flag at the right moment. Gibson also make several other games on similar lines one of which, 'Tri-tactics', involves sea, land and air forces and in which the capture of naval bases is a means of winning.

The best-known manufacturer of board war games is undoubtedly the Avalon Hill Company of Baltimore, and at one time they had four naval games in production. Three of these —'U-boat', 'Bismarck' and 'Jutland'—have now been withdrawn, and in the case of 'Jutland' this is particularly disappointing as it was a well-designed game.

The only Avalon Hill game now in production is 'Midway', an attractively presented and packaged version of the great naval battle of June 1942 in which the United States wrested air supremacy from the Japanese in the Pacific War. This battle occurred when several Japanese fleets converged on Midway Island with a view to bringing the American fleet to what it was hoped would be decisive combat. In fact the very much smaller American force had the advantage of knowing their enemy's intentions and approximate movements and this factor, in part, enabled them to sink the enemy's principal aircraft carriers which in turn obliged the Japanese battleships, cruisers and landing ships to retreat.

American superiority in the matter of knowledge of Japanese movements is built into the game by giving the American player more opportunities each turn to locate the enemy ships. It is the successful exploitation of this advantage which generally enables the American player to make decisive strikes against the first and second of the Japanese forces to appear on the game board and thus win the entire game. The search aspect of the game is virtually only that of the 'Battleship' games. Each player moves his ships secretly and in turn asks the opponent if his ships are on particular squares. If combat occurs it is resolved in a somewhat over-simplified fashion on a separate battle board. The makers recommend that the basic rules be mastered first and then in the Tournament Game more sophisticated rules are introduced.

The initial situation for each side is intriguing. The Japanese player must reach and invade Midway Island within a specified number of game moves, but his great armada appears on the movement board only in four groups, widely spaced in time. Handicapped in the search procedure, he must do his best to escape discovery in the early moves until his overwhelming superiority in ships can become effective in combat. In contrast, the American player has all his ships on the movement board from the outset, and he ought to be able to discover the enemy and get in an air strike quickly whilst the odds still favour him. But if his freedom of manoeuvre does not lead to him getting in a telling blow early on, his opponent will be bound to overwhelm him eventually.

Altogether 'Midway' remains one of my favourites (even if

realism has been sacrificed for playability in several important elements of the rules) since it gives both players reasonable scope for employing different tactics each time it is played.

COMPLEX GAMES

'Complex' will be taken by some readers as a compliment to the games so described here or as a criticism. On balance I use it in the complimentary sense as meaning those games which achieve a high degree of realism although often at the expense of time, and sometimes at the expense of playability.

The first game of this type is 'Surface Warship' which is distributed by the publishers of *The Courier*, a New England war-game journal. It is a modestly priced game and would warrant inclusion in Chapter 3, since it can be used with model ships, were it not for the fact that it comes as a complete game in itself with rules, cardboard deck plans and other items. The rules are subdivided into 'basic', 'advanced' and 'optional' categories with the latter providing for such niceties as speed and depth settings for torpedoes, the use of star shells and searchlights in night actions, and glide and dive-bombing runs for aircraft. The two situations for which the rules are especially intended are the Battle of the River Plate and *Bismarck*'s encounter with *Hood* and *Prince of Wales*. Although the rules owe something to those of Fletcher Pratt, they are modified to provide for a fast-moving game in which each player commands a ship and in which every move must take only thirty seconds. Players have to react quickly to changed situations, and combat results depend on visual accuracy and the correct selection of ranges rather than on dice throws.

The final trio of games dealt with at length in this chapter are all the work of Simulations Publications Incorporated (SPI) of New York and are distributed within copies of their magazine *Strategy and Tactics* or as boxed games. In fact, subscribers to the magazine get at least one war game included with every issue, although the land games have so far greatly outnumbered all others so that the naval enthusiast may prefer to buy only those issues containing a naval game. Clearly, a game issued between the pages of a magazine cannot have a stiff game board, but each one does have a folding sheet-map which opens out to a substantial size and can then be given a stiff backing by the purchaser. The other playing pieces are cardboard and press out from sheets in the manner of Avalon Hill products. Naturally, a great many games can be promoted in this fashion and the manufacturers are not obliged to confine their efforts to marketing a handful of proven best-sellers, as is the case with firms producing expensive boxed games. It seems likely that we shall see more of such distribution-by-magazine. Indeed, a new magazine, *Battle Flag*, launched in the autumn of 1973, plans to include a game with every other issue.

To return to the games issued by SPI: 'The Flight of the Goeben' may be played in several versions—as a basic or advanced strategy game, or as a tactical exercise. Moreover, victory conditions for each side are chosen in a random fashion at the beginning of the game and neither player knows who has won until points are added up at the end. This apparently haphazard state of affairs mirrors the situation when, in August 1914, the very fast German warships *Goeben* and *Breslau* were in the central Mediterranean and could be expected to make a dash for the Atlantic or attack British and French convoys in the Mediterranean. In the event, they received orders to proceed to Constantinople where it was hoped the pro-German government would give the ships a secure base for future operations. So British and French warships, all of them slower than their German counterparts, had to be deployed for a number of eventualities.

The rules give particular prominence to fuel consumption, and the Allies have access to friendly ports whilst the German ships can meet up with colliers. The British ships start the game at Malta, the French at Toulon, and the German cruisers can be either in the Tyrrhenian Sea or the Ionian Sea. Should Austria and Italy join in, then their home bases are Pola and

Messina respectively. As well as achieving their general objective of escaping from the Allies, the German ships can amass points by sinking enemy warships and freighters and by shelling enemy towns. Search procedures dominate this game and so the board is divided into seas, for example the Adriatic Sea, which are subdivided into smaller areas, which in turn consist of numbered hexagons. A player commences to search by asking if the enemy has ships in the seas in which he himself has vessels and then uses supplementary questions which may or may not reveal the enemy's precise strength and exact location. In the strategic game combat is a simple affair, but if the tactical rules are employed then a more realistic encounter occurs although with a consequent slowing down in the game's pace.

In 1971 SPI issued a more ambitious game, 'USN', which allows for nothing less than the waging of naval war throughout the Pacific theatre of war between 1941 and 1943. The large map provided has as its parameters the Aleutian Islands, the Hawaiian Islands, New Zealand and Burma and is in itself a useful piece of equipment for the model-ship user who could press the map into service for the movement side of his Pacific actions. The playing pieces depict all the ships which were available to the USA and Japan between 1941 and 1943, together with land units, air bases and transport units. The rules are lengthy and complicated, and the user is strongly recommended to play through the four mini-games. Since the mini-games deal with the battles of the Coral Sea, Midway, the Eastern Solomons and the Santa Cruz Islands it can be readily seen that the full-scale game of 'USN' is a considerable undertaking! The novice may also practise for this ultimate game by playing through two suggested campaign games as well as the mini-games. In all six of these situations, points are awarded for the destruction of enemy units and for the securing of supremacy in particular zones of the Pacific. When one side achieves the total of points matching its victory conditions then the game is over.

'USN' goes much further than any other board game in

providing rules and playing pieces which enable the combatants to play an almost unlimited number of scenarios with a high degree of realism. The cost is in the time and patience needed to master just some of the rules and then to play through a move so that 'USN' must be adjudged a flexible vehicle for many sorts of naval operation rather than as a playable game, complete in itself. Dennis Brackman of West Pymble, Australia, who must be one of the keenest of all naval board gamers, has kindly written to me at length of his experiences with all three SPI games. Of 'USN' he writes, 'I have played "USN" many times . . . and enjoy it to the exclusion of all others. The only problem is finding another player who enjoys extreme complication and protracted planning.'

The third SPI game to appear was 'CA' which was published in 1973. The scene is again the Pacific War, but this time the game chart is only of the seas around the eastern end of the Solomon Islands. The ships available to the players are all surface vessels from battleships down to destroyers, but excluding aircraft carriers. Ten scenarios are suggested as games, each resembling a situation pertaining in these waters between 1941 and 1945. 'CA' is thus a game of tactical naval warfare in which victory conditions are either the sinking of the greatest number of ships or the denying of a route to an enemy trying to get at a convoy or other vulnerable target. Since all movement and combat takes place on the same playing area, there is neither secrecy about movement nor the identity of warships. Instead, the designers of the game have tried to give proper emphasis to the differences in performance of, for example, American and Japanese torpedoes, differences in the speeds of warships, and the Japanese panache for night fighting. A grand game, then, for those who like to see battleships and cruisers getting full scope to show their paces with never an aircraft carrier in sight to spoil things!

CONCLUSION

We have come a long way from 'Battleships' to 'CA' and yet there is still surely plenty of room for more naval games of the standard and complex variety. But the situation in the immediate future does not look too hopeful. Avalon Hill retain 'Midway' in their lists, but apart from their 'Leyte Gulf' variant published in the October 1973 issue of *The General* they do not seem to have plans for any other modern naval games. I do not know the fate of Western Enterprises of Hales Corner, Wisconsin, who in 1969 were publishing game kits for the Battle of the Falklands (1914) and for the final *Scharnhorst* action (1943) and had plans for many more. On the other hand, SPI do intend to publish 'Leyte Gulf' as a game in an issue of *Strategy and Tactics* and in Britain, Tony Jones of East Preston, Sussex has designed two games of the World War II period 'Wolf Pack' and 'PQ17' which he hopes will be marketed. His air game 'Schweinfurt' was well received on its publication some years back.

Finally, and also on the credit side, three fine magazines—*The General* and *Strategy and Tactics* in the United States and *Albion* in Britain—keep the board war gamer closely in touch with developments. I have also noticed a tendency for an increasing number of more general British magazines to carry articles about commercial war games, written by acknowledged experts in the field. For example, Donald Turnbull, editor of *Albion*, has contributed to *Games and Puzzles* and John Merrills of Sheffield to *Military Modelling*, so that the subject is reaching a far wider audience than that of the more specialised magazines and some more recruits for naval gaming may result.

NOTES

Chapter 1

The following firms manufacture and distribute some of the ships described in this chapter.

ALNAVCO: Box 9, Belle Haven, Virginia 23306, USA. Catalogue available. For all Superior ships, complete or in cheaper, near-finished form.

FLEETLINE: 2 Lees Lane, Gosport, Hampshire, England.

GILLOW SHIPS: R. Gillow, 62 Bishopton Road, Stockton-on-Tees, Tees-side, England.

LEICESTER MICROMODELS: 5 Goss Barton, Nailsea, Bristol, England.

MINIFIGS: Miniature Figurines Ltd, 28/32 Northam Road, Southampton SO2 0PA, Hampshire, England.

MODEL FIGURES AND HOBBIES: Lower Balloo Road, Groomsport, Co Down, Northern Ireland. For Ensign ships and H/R accessories.

NAVAL WAR GAMES SOCIETY (UK): 16 Hugo Road, London N19 5EU, England. For all Navwar 1/3,000 scale ships.

OCEANIC: F. G. Skinner, Hotel Continental, St Margaret's Road, Torquay, Devon, England. Mr Skinner's hotel bar has permanent displays of his Oceanic range together with 14,500 other ships, all to 1/1,250 scale.

SKYTREX: 28 Church Street, Wymeswold, Leicestershire, England.

SUPPLIERS OF MODELS MENTIONED IN THIS CHAPTER

(Dealers appreciate the enclosure of a stamped addressed

envelope or where appropriate an International Reply Coupon with initial enquiries, requests for catalogues etc.)

UNITED KINGDOM

ALMARK MODEL PRODUCTS: 270 Burlington Road, New Malden, Surrey. A main distributor for Casadio Miniships.

ARGYLE MODELS: 247 Argyle Street, Glasgow G2 8DN and 103 High Street, Edinburgh EH1 1SG, Scotland. For Delphin, Hansa, Navis, Saito, Trident, Wiking models.

THE ARMOURY: 1 Fisher Road, Stoke, Plymouth PL2 3BA. For Leicester Micromodels. Hoping to produce own range from late 1974.

HOLTON, JOHN: 2 Aragon Avenue, East Ewell, Epsom, Surrey. Supplies all the Superior range and generally has a few new or second-hand models by other firms, including Eaglewall, on his special offer list.

MAGPIE MODELS: 146 Harrogate Road, Leeds 7. For Leicester Micromodels.

MODELS FIGURES AND HOBBIES: address already given. For Fleetline and Triang ships.

SKYTREX: address already given. All Navwar products.

1,200 MODEL SHIP SOCIETY: 18 Wroxton Road, Nunhead, London, SE15. Gives a discount to ship purchasers who have an annual subscription for society membership and take the journal. Holds stocks of the German producers in 1/1,200 and 1/1,250 scale and Casadio Miniships.

FEDERAL REPUBLIC OF GERMANY

HOFFMAN shops of Berlin. For most German makes and Oceanic models.

NATHAN R. PRESTON: Preston Hobby Modelle OHG, PO Box 2388, 86 Bamberg. For most continental makes of model.

SCHMIDT, CHRISTIAN: 8 München 55, Sauerbruchstrasse 10. For Copy, Delphin, Hansa, Mercator, Neptun, Star, Trident, Wiking.

WIEDLING, EGON: 8 München 2, Theatinerstrasse 13. For Anker,

Atlantic, Copy, Delphin, Hansa, Mercator, Mercury, Miniships, Navis, Neptun, Oceanic, Schlingelhof, Star, Triang, Trident, Wiking—even Eaglewall occasionally.

UNITED STATES OF AMERICA

COKERCRAFT: PO Box 124, Charleston, South Carolina. All the German models and plastic kits.

MINIATURE FIGURINES USA INC: 4311 Lemon Avenue, Dallas, Texas 75080. For all British Skytrex products.

MPC: 126 Groesbeck Highway, Mt Clemens, Michigan 48043. For Casadio Miniships.

SPARTANS INTERNATIONAL: 4121 Long Beach Boulevard, Long Beach, California 90807. Expecting to distribute their own range of plastic resin ships in 1/1,200 scale.

VALLEY PLAZA HOBBIES: 12160 Hamlin Street, North Hollywood, California 91606. Mainly stocks Navis ships.

WHAT THE MODELS COST

Prices cannot really be meaningful, since three major factors invalidate the information that might be put down. First, the effects of inflation mean that prices are constantly subject to increase. Second, the frequent fluctuations in exchange rates involving the three countries most concerned in either the manufacture or the distribution of the models—Britain, Germany and the United States—has a similarly confusing effect. Third, many of these models can be bought in several ways—over the counter, by mail order or by mail order from abroad.

However, the models can be grouped in the following broad categories. The prospective purchaser will, of course, enquire of the dealer or supplier of his choice.

Very inexpensive

Airfix, Eaglewall, Ensign, Gillow, Leicester Micromodels, Lindberg, Minifigs, Miniships, Navwar, Pyro, Saito, Skytrex, Triang.

Inexpensive

Recognition Models, Schlingelhof, Star, Wiking.

Medium

Copy, Delphin, Fleetline, Mercator, Mercury, Navis, Superior.

Expensive

Anker, Hansa, Neptun, Oceanic, Trident.

Chapter 2

I do not know of any book which deals with the whole subject of building waterline ship models for the modern period. But the following will all prove useful for different aspects of the subject.

Bowen, J. *Waterline Ship Models* (London, 1972)
Craine, J. H. *Ship Modelling Hints and Tips* (London, 1973)
Deason, G. H. *Cardboard Ship Models* (Hemel Hempstead, 1972)
Ellis, C. *How to Go Advanced Plastic Modelling* (Hemel Hempstead, 1970)
——. *How to Go Plastic Modelling* (Hemel Hempstead, 1973)
—— (ed). *Airfix Magazine Annual for Modellers* (London, 1973)
Grimwood, V. R. *American Ship Models and How to Build Them* (New York, 1942)
Hodges, P. *Royal Navy Warships' Camouflage* (London, 1973)
Hutchinson, J. K. D. *The Model Ship Builder* (London, 1949)
(no author) *Solarbo Book of Balsa Models* (London, 1969)
Warring, R. H. *Modelling with Balsa* (London, 1967)
Williams, G. R. *The World of Model Ships and Boats* (London, 1971)

Scale plans for twentieth-century warships are produced by A and A Plans (London), 102 Mattison Road, London N4, and other agents stocking them are:

ACTION BOOKS: PO Box 5611, Buena Park, California 90620, USA.

HERSANT, W. E.: 228 Archway Road, London N6.
SEAGULL MODEL (GB): PO Box 62, London SE 19.

Warship accessories in 1/1,200 or 1/1,250 scale are distributed by Alnavco, Fleetline, Hansa and Model Figures and Hobbies. See Chapter 1 for details.

Chapter 3

The rules of Fred Jane and Fletcher Pratt will be found in the following books.

Jane, Fred T. *Hints on Playing the Jane Naval War Game* (London, 1901)

——. *How to Play the Naval War Game* (London, 1912)

Pratt, Fletcher. *Fletcher Pratt's Naval War Game* (New York, 1940), reprinted in 1974 by D. Darcy, 1841 South Halsted, Chicago, Illinois 60608

LEICESTER MICROMODELS: 5 Goss Barton, Nailsea, Bristol, England, distribute *The Last Sea Battles* (1914–18) and *Naval Warfare* (1939–45). Their author, Keith Robinson, is pleased to answer comments on them at 50 Walcot Walk, Peterborough, Northamptonshire.

NAVWAR: 16 Hugo Road, London N19, England, distribute the the World War II rules of the Naval War Games Society (UK).

SEAPOWER: address in notes for Chapter 1, under Alnavco.

SKYTREX: address in notes for Chapter 1. The author of these World War I rules is John Hammond.

Victory at Sea from The Nassau Naval and Military Historical Society, Box 53, New Hyde Park, New York 11040, USA.

Chapter 5

For constructing accurate charts for campaign purposes, and specifically for making the movement chart for the East Asian campaign described in this chapter, I have found the *Pergamon*

World Atlas (Oxford, 1968) most useful. A particularly helpful feature is that it provides in nautical miles the distances between all ports for a given area.

Chapter 6

BOOKS

Boyd, Norman. *Discovering Ship Models* (Tring, Hertfordshire, 1971)

Clerk, John. *An Essay on Naval Tactics* (Edinburgh, 1790). See also the 1804 and 1827 editions

Dunn, P. *Sea Battle Games* (London, 1970)

Featherstone, Donald E. *Naval War Games* (London, 1965)

Jane, Fred T. *Hints on Playing the Jane Naval War Game* (London, 1901)

——. *How to Play the Naval War Game* (London, 1912)

Pratt, Fletcher. *Fletcher Pratt's Naval War Game* (New York, 1940), reprinted in 1974 by D. Darcy, 1841 South Halsted, Chicago, Illinois 60608

Taylor, A. *Rules for Wargaming* (Tring, Hertfordshire, 1971)

Wilson, Andrew. *War Gaming* (London, 1970). (Previously published as *The Bomb and the Computer*, 1968)

JOURNALS

Alnavco Log, quarterly. $2.00, USA. Alnavco Naval Division, Box 9, Belle Haven, Virginia 23306, USA.

Battlefleet, monthly, £1.10, UK. $5.00, USA air mail. The Naval Wargames Society (UK), 16 Hugo Road, London N19 5EU.

Broadside, quarterly. £1.30, UK. $6.00, USA air mail. 1,200 Model Ship Society, 18 Wroxton Road, Nunhead, London SE15 2BN.

The Courier, eight issues each year. A magazine about warfare with miniatures and board games. $4.00 a year USA. $5.00 a year foreign, from 45 Willow Street, Brockton, Massachusetts 02401, USA.

Military Modelling, monthly, from all newsagents, or by

subscription £2.35 ($6 for overseas subscribers) from Model and Allied Publications Ltd, 13–35 Bridge Street, Hemel Hempstead, Hertfordshire.

Miniature Warfare and Model Soldiers, monthly. £3.15 ($10.50) annual subscription from Stanhope House, Fairbridge Road, London N19 3HZ.

Signal, fortnightly. Ten issues for £1 in United Kingdom; six issues for $1, or twenty for $3 in North America; ten issues for 5DM on the continent. North American subscriptions to *Signal*, PO Box 830, CFPO 5056, Ontario, Canada; European subscriptions to *Signal*, PO Box 830, CFPO 5056, 757 Baden-Baden 1, Germany.

Swabbers, $2 for a year's subscription from H. Barents, 157 State Street, Zeeland, Michigan, USA.

Wargamers' Newsletter, monthly, £3.60 (£3.85 overseas and $11.00 in USA) from Tradition, 188 Piccadilly, London W1V 4DA.

Warship International, quarterly, from $6 in the USA where it is published at Box 3249, First Street Station, Radford, Virginia 24141. Canadian agent: K. MacPherson, 79 Aldershot Crescent, Willowdale, Ontario; UK agent: A. Gilbert, 9 Cranbrook, Camden Street, London NW1 0LJ.

A warning on prices! It will be readily appreciated that the subscription rates given here are likely to change, and generally to increase, since printing and postal costs have increased frequently in recent years.

Chapter 7

The games mentioned in this chapter are made by the following companies:

'Admirals' Pallitoys, Parker's, Coalville, Leicester, England.

'Bataille Navale' CEJI, Département Jouets Educatifs Universels, 14 à 20 rue Bon-Houdart, 93700 Drancy (BP 19), France.

'Battleships' Merit Games, J. and L. Randall, Potters Bar, Hertfordshire, England.

'Battleships' J. Waddington Ltd, Woodlesford, Leeds, England.

'Dover Patrol' H. P. Gibson and Sons Ltd, 312 Kingston Road, London SW20, England.

'Flight of the *Goeben*', 'USN', 'CA' Simulations Publications Inc, 44 East 23rd Street, New York, NY 10010, USA.

'Les Grands Amiraux' Miro Company (Editions Capiepa), 27 avenue Pierre-1er-de-Serbie, 75116, Paris, France.

'Midway' The Avalon Hill Company, 4517 Harford Road, Baltimore, Maryland 21214, USA.

'Surface Warship' The Courier, 45 Willow Street, Brockton, Massachusetts 02401, USA.

Of these games, the following are all readily available from hobby shops and department stores in the country of origin: 'Admirals', both 'Battleships', 'Dover Patrol' (Britain); 'Bataille Navale', 'Les Grands Amiraux' (France); 'Midway' (USA).

'Surface Warship' is only available from the address given above. SPI titles are available from the New York office but can also be obtained from:

PO Box 46, Altrincham, Cheshire, England.

Skytrex Ltd, 28 Church Street, Wymeswold, Leicestershire, England.

North Shore War Games, 57 Wallalong Crescent, West Pymble, NSW 2073, Australia.

'Midway' is also available from:

North Shore War Games (address as above, Australia).

James Block, 5 Wurth Place, Chifley 2606, Canberra, Australia.

In England the game is available from:

Avalon Hill Games, 646 High Road, Finchley, London.
Ashley Brown, 44 Linksway, Northwood, Middlesex.
The General Trading Co, 144 Sloane Street, London.
Hamley's Toy Shop, 200–202 Regent Street, London.
Harrod's Store, Knightsbridge, London.
M. E. Maitland, 243 Regent Street, London.
Soldiers, 36 Kennington Road, London.
Tradition, 188 Piccadilly, London.

Battle Flag magazine and games are published at 465 Woodland Hills, Philadelphia, Mississippi 39350, USA.

APPENDIX

AUSTRALIA AND NEW ZEALAND

War gaming in general has enjoyed a recent upsurge in popularity in Australia due in part to more favourable rates of exchange with Britain and the United States which has made many of their products more attractive in price. Hobby shops are also stocking more war-game items both for the board-game player and the miniatures' enthusiast. A bi-monthly publication *Miniature Manoeuvres* has a large readership, and includes in each issue a page giving war-game clubs and their addresses. It is published at 50 Clissold Parade, Campsie, NSW 2194. Amongst the several clubs operating in Australia are two particularly large and well-established ones:

NSW Wargames Confederation, PO Box 249, Guildford, NSW 2161.

Victorian Wargames Association, 1/182 Dorcas Street, South Melbourne, Victoria 3205.

New Zealand's naval enthusiasts seem to give pride of place to the Napoleonic era but Fletcher Pratt-style World War II games using large-scale Airfix ships have been played occasionally at some clubs. But battles in the modern period with 1/3,000 or 1/4,800 scale ships are on the increase and Leicester Micromodels, Skytrex and Navwar products are thus widely used. Up to the present time, board-game enthusiasts have had to send abroad for their sets.

War gamers contact each other mainly through placing advertisements in model shops and general hobby journals such as the British *Military Modelling*, and they can also meet at conventions and exhibitions.

The following are well-established clubs:

Auckland Wargames Society: Gavin Kennedy, 84 Roseberry Avenue, Birkenhead, Auckland 10.

The Otago Miniature Tacticians Society, PO Box 5474, Dunedin.

The Warlords: V. J. Thomas, 53 Lyndhurst Street, Palmerston North.

Wellington Wargames Section: P. Graham, 24 Tweed Road, Papakowhai, Paremata, Wellington.

Wellington Wargames Society Inc: M. H. S. Stevens, Weir House, Gladstone Terrace, Wellington.

GREAT BRITAIN

Although still the poor relation of land war gaming, naval gaming is decidedly on the increase and for the ancient, sailing-ship and modern periods there is a ready supply of ships and rules. In Chapter 1 we saw just how many firms are producing ships for the modern period in 1/1,200 and 1/1,250 scale, but I suspect that we shall now see the scales of 1/3,000 and 1/4,800 coming to the fore and winning many new adherents to the hobby. Board games are becoming increasingly available 'over the counter', but mail order is still the mainstay, as in all other aspects of the hobby. Board war gamers are well served by the bi-monthly journal *Albion* which is edited by D. Turnbull, Flat 13, Gilmerton Court, Trumpington Road, Cambridge, although its continuance is in some doubt.

Many opportunities exist for war gamers to contact each other, and the motorway system and inter-city rail services enable the two London-based naval war game societies to claim reasonably national memberships. They are the Naval War Games Society and the 1,200 Model Ship Society and their activities, journals and services have been described in Chapter 6.

THE UNITED STATES AND CANADA

Despite the great distances which often separate one group of war gamers from another in the United States, the hobby has reached a high degree of organisation in many respects and the frequent conventions which occur give even the most isolated war gamer the opportunity to sally out and make contacts with fellow players. But the hobby is strongest in the urban centres and here clubs or even federations of clubs cater for the keen player. The largest of these groups is Spartan International Inc, whose president is Russell Powell and which has its headquarters at Box 1017, Bellflower, California 90706. Spartan has some 1,000 members who compete for cash or for trophies in a whole range of war-game periods with both miniatures and with board games. Standardisation of rules has thus become a necessity, and the modern naval ones are based on an original game by Ivan Travniceck. A new set by A. D. Zimm—*Battle Stations*—which may be adopted for future Spartan naval contests appeared in full in the company journal *The Spartan* in issues five and six of 1973. A corporate feeling is fostered in Spartan International by the wearing of uniform and the assigning of ranks, but the company is also found supporting the hobby in general in numerous ways, most notably by sponsoring conventions in various parts of the United States. Yet another service provided by Spartan is the production of a monthly news sheet *Gamers Guide* which lists war-game clubs state by state and has advertisements from war gamers seeking opponents, announces new products, and reviews the various hobby journals.

On a more modest scale the International Federation of Wargaming serves the state of Illinois but invites membership from gamers in other states. Its headquarters is c/o Len Lakofka, 4970 Marine Drive, Apt 525, Chicago, Illinois.

The United States is the home of board war gaming. Amongst the many magazines which report on it are the widely read *The General* (Avalon Hill), *Strategy and Tactics* (SPI) and *Conflict* (Simulation Design Incorporated).

Canadian naval war gamers are strongest on the ground in Vancouver but keep in touch nationally through *The Signal* (see Chapter 6) and *The Trumpeter* (editor Jack Hutchings), 4578 Bentlawn Avenue, Burnaby 2, British Columbia.

BIBLIOGRAPHY

There is a very substantial literature for both World Wars. The reader will, of course, find many more reading suggestions in each of the books he tackles, but three bibliographies which have been of particular value to me are:

Albion, R. G. *Naval and Maritime History: An Annotated Bibliography* (Mystic, Conn, 1972 and Newton Abbot, 1973)

Department of the Navy. *United States Naval History—a Bibliography* (Washington DC, 1972)

Ziegler, J. *World War II—Books in English, 1945–65* (Stanford, Calif, 1971)

WORLD WAR I

Corbett, J. S. and Newbolt, H. *Naval Operations*, 5 vols (London, 1920–38)

Dupuy, T. N. *The Military History of World War I*, vols 9–10 (New York, 1967)

Hoehling, A. A. *The Great War at Sea* (New York, 1965)

Marder, A. J. *From the Dreadnought to Scapa Flow 1904–19*, 5 vols (London, 1961–70)

Purnell's History of the First World War, 8 vols (Bristol, 1969)

WORLD WAR II

Morison, S. E. *History of the United States Naval Operations in World War II*, 15 vols (Boston, 1947–62)

——. *The Two-Ocean War: A Short History of the United States Navy in the Second World War* (Boston, 1963)

Purnell's History of the Second World War, 8 vols (Bristol, 1966)

Roskill, S. W. *The War at Sea 1939–45*, 3 vols (London, 1954–60)

WARSHIP DATA

Four excellent sources for ship data, plans and photographs are:

Jane's Fighting Ships All the following have been reprinted by David & Charles (Newton Abbot): 1898, 1905/6, 1906/7, 1914, 1918, 1919, 1924, 1931, 1939, 1944.

Ian Allan (London) *Warships of World War I* and *Warships of World War II*. These are on a country-by-country basis.

Almark (London) A new and growing series for World War II. Especially good because of the scale plans is Blundell, W. D. G. *Royal Navy Warships 1939–45* (London, 1971).

Macdonald (London) *Navies of the Second World War*. Again, country-by-country but subdivided into ship categories for very large navies.

ACKNOWLEDGEMENTS

A great many people have helped me with this book and I hope that they will feel that the finished product is an adequate tribute to their kindness. Certainly it has been both a pleasure and an encouragement to correspond with so many people who share a common interest in naval history and naval war games.

Particular thanks are due to:

Michael F. Childs, Lecturer in Cartography at Oxford Polytechnic, who responded enthusiastically to my request for help with diagrams and maps. As well as producing the final line drawings, he contributed a great deal to their design. Roughs for his three maps were drawn by Diploma students David Lewis and Pieter Summerfield.

Iradj Parvaneh, also of Oxford Polytechnic, who brought the same careful attention to the preparation and execution of the photographs.

Anthony Bowditch of Saltash, Cornwall, a valued friend and war-game opponent for many years, who read the manuscript at several stages and made helpful suggestions for its improvement.

I am also very grateful to the following war gamers who gave generously of their time and knowledge to enable me to describe various aspects of the hobby, especially for their own countries: Dennis Brackman, West Pymble, Australia; Charles Duggie, Riverside, California, USA; Wayne Hoheisel, Greenbelt, Maryland, USA; Damian Housman, Apple Valley, California, USA; Tony Jones, East Preston, Sussex; John Mansfield, the Canadian armed services; Donald Miller, Wheaton, Maryland, USA; Mark Stevens, Wellington, New Zealand and Donald Turnbull, Cambridge.

ACKNOWLEDGEMENTS

For assistance with Chapter 1 I thank John Hammond of Skytrex Limited; Model Figures and Hobbies of Groomsport, Northern Ireland; Peter Paschall of Alnavco Incorporated; F. Gerald Skinner of Torquay, Devon and Dennis Teague of Plymouth, Devon.

I thank the following for assistance with Chapter 7: Richard Bryant, Brockton, Massachusetts, USA; Mario Cantinelli, Hyères, France; John Merrills, Sheffield, Yorkshire; M. H. Nethercott, Ingatestone, Essex and Robert Stuart, Orpington, Kent.

Finally, I thank my wife Sylvia who had the tedious task of typing early drafts and who made many valuable suggestions for improving the clarity of the text. Her constructive criticisms coupled with her sustained support for the whole project proved the ideal spur to me as I frequently had need of both.

INDEX

(*figures in italics denote illustrations*)